Advanced Structured Materials

Volume 189

Common engineering materials are reaching their limits in many applications, and new developments are required to meet the increasing demands on engineering materials. The performance of materials can be improved by combining different materials to achieve better properties than with a single constituent, or by shaping the material or constituents into a specific structure. The interaction between material and structure can occur at different length scales, such as the micro, meso, or macro scale, and offers potential applications in very different fields.

This book series addresses the fundamental relationships between materials and their structure on overall properties (e.g., mechanical, thermal, chemical, electrical, or magnetic properties, etc.). Experimental data and procedures are presented, as well as methods for modeling structures and materials using numerical and analytical approaches. In addition, the series shows how these materials engineering and design processes are implemented and how new technologies can be used to optimize materials and processes.

Advanced Structured Materials is indexed in Google Scholar and Scopus.

Andreas Öchsner · Resam Makvandi

A Numerical Approach to the Classical Laminate Theory of Composite Materials

The Composite Laminate Analysis Tool—CLAT v2.0

Springer

Andreas Öchsner
Faculty of Mechanical and Systems Engineering
Esslingen University of Applied Sciences
Esslingen am Neckar, Baden-Württemberg
Germany

Resam Makvandi
Acandis GmbH
Stuttgart, Baden-Württemberg, Germany

ISSN 1869-8433 ISSN 1869-8441 (electronic)
Advanced Structured Materials
ISBN 978-3-031-32974-6 ISBN 978-3-031-32975-3 (eBook)
https://doi.org/10.1007/978-3-031-32975-3

This Springer imprint is published by the registered company Springer Nature Switzerland AG
The registered company address is: Gewerbestrasse 11, 6330 Cham, Switzerland

Preface

Composite materials, especially fiber-reinforced composites, are gaining increasing importance since they can overcome the limits of many structures based on classical engineering materials. Particularly, the combination of a matrix with fibers provides far better properties than the single constituents alone. Despite their importance, many engineering degree programs do not treat the mechanical behavior of this class of advanced structured materials in detail. Thus, some engineers are not able to thoroughly apply and introduce these modern engineering materials in their design process.

This monograph first provides a systematic and thorough introduction to the classical laminate theory based on the theory for plane elasticity elements and classical (shear-rigid) plate elements. The focus is on unidirectional lamina which can be described based on orthotropic constitutive equations and their composition to layered laminates. In addition to the elastic behavior, failure is investigated based on the maximum stress, maximum strain, Tsai-Hill, and the Tsai-Wu criteria.

The solution of the fundamental equations of the classical laminate theory is connected with extensive matrix operations and many problems require in addition to iteration loops. Thus, a classical hand calculation of related problems is extremely time-consuming. In order to facilitate the application of the classical laminate theory, we decided to provide a Python-based computational tool, the so-called Composite Laminate Analysis Tool (CLAT) to easily solve some standard questions from the context of fiber-reinforced composites. The tool runs in any standard web browser and offers a user-friendly interface with many post-processing options. The functionality comprises stress and strain analysis of lamina and laminates, derivation of off-axis elastic properties of lamina and the failure analysis based on different criteria.

We look forward to receiving some comments and suggestions for the next release of CLAT and edition of this textbook.

Esslingen am Neckar, Germany — Andreas Öchsner
Stuttgart, Germany — Resam Makvandi
March 2023

Contents

Symbols and Abbreviations

Latin Symbols (Capital Letters)

A	Area, cross-sectional area
$\boldsymbol{A}$	Submatrix of the generalized elasticity matrix
$\boldsymbol{A}'$	Inverted submatrix
$\boldsymbol{B}$	Submatrix of the generalized elasticity matrix
$\boldsymbol{B}'$	Inverted submatrix
C	Element of elasticity matrix
$\overline{C}$	Element of transformed elasticity matrix
$\boldsymbol{C}$	Elasticity matrix
$\overline{\boldsymbol{C}}$	Transformed elasticity matrix
$\boldsymbol{C}^*$	Generalized elasticity matrix
D	Element of the compliance matrix
$\overline{D}$	Element of transformed compliance matrix
$\boldsymbol{D}$	Compliance matrix, submatrix of the generalized elasticity matrix
$\boldsymbol{D}'$	Inverted submatrix
$\overline{\boldsymbol{D}}$	Transformed compliance matrix
E	Elasticity modulus
F	Force
G	Shear modulus
I	Second moment of area
M	Moment
$\boldsymbol{M}$	Column matrix of moments
N	Normal force (internal)
$\boldsymbol{N}$	Column matrix of normal force
Q	Shear force (internal)
$\boldsymbol{Q}$	Column matrix of shear forces
R	Strength ratio
$\boldsymbol{R}$	Transformation matrix
$\boldsymbol{T}$	Transformation matrix
V	Volume

Latin Symbols (Small Letters)

a Geometric dimension, coefficient
b Geometric dimension
$\boldsymbol{b}$ Column matrix of body forces
$\boldsymbol{e}$ Column matrix of generalized strains
f Body force
$\boldsymbol{g}$ Standard gravity
$k_{1\mathrm{c}}$ Longitudinal compressive strength in direction *1*
$k_{1\mathrm{t}}$ Longitudinal tensile strength in direction *1*
$k_{12\mathrm{s}}$ In-plane shear strength in plane *12*
$k_{2\mathrm{c}}$ Longitudinal compressive strength in direction *2*
$k_{2\mathrm{t}}$ Longitudinal tensile strength in direction *2*
$k^{\varepsilon}_{1\mathrm{c}}$ Longitudinal compressive failure strain in direction *1*
$k^{\varepsilon}_{1\mathrm{t}}$ Longitudinal tensile failure strain in direction *1*
$k^{\varepsilon}_{12\mathrm{s}}$ In-plane shear failure strain in plane *12*
$k^{\varepsilon}_{2\mathrm{c}}$ Longitudinal compressive failure strain in direction 2
$k^{\varepsilon}_{2\mathrm{t}}$ Longitudinal tensile failure strain in direction *2*
m Distributed moment, mass
n Layer number
q Distributed load in thickness direction, area-specific load
$\boldsymbol{q}$ Column matrix of distributed loads
s Column matrix of stress resultants
t Lamina or laminate thickness
u Displacement
$\boldsymbol{u}$ Column matrix of displacements
x Cartesian coordinate
y Cartesian coordinate
z Cartesian coordinate

Greek Symbols (Small Letters)

α Rotation angle
γ Shear strain (engineering definition)
ε Normal strain
$\boldsymbol{\varepsilon}$ Column matrix of strain components
η Shear-extension coupling corefficient
κ Curvature
$\boldsymbol{\kappa}$ Column matrix of curvature components
ν Poisson's ratio
ϱ Mass density
σ Stress

$\boldsymbol{\sigma}$ Column matrix of stress components
φ Rotation angle

Mathematical Symbols

$\mathcal{L}\{\dots\}$ Differential operator
$\boldsymbol{\mathcal{L}}$ Matrix of differential operators

Indices, Superscripted

$\dots^{\mathrm{n}}$ Normalized
$\dots^{0}$ Related to the middle-surface

Indices, Subscripted

$\dots_{1}$ Direction *1*
$\dots_{2}$ Direction *2*
$\dots_{12}$ Plane *12*
$\dots_{\mathrm{c}}$ Compression
$\dots_{\mathrm{s}}$ Shear, symmetric
$\dots_{\mathrm{t}}$ Tension

Abbreviations

BC Boundary condition
BEM Boundary element method
CLT Classical laminate theory
FDM Finite difference method
FEM Finite element method
FVM Finite volume method

Chapter 1
Introduction

Abstract This chapter introduces the major concept of composite material, i.e., the combination of different components, to obtain in total much better properties than a single component for itself. The focus is on fiber-reinforced composites where a single layer has unidirectionally aligned reinforcing fibers. The chapter concludes with a brief literature review related to textbooks which cover the mechanical behavior of fiber-reinforced materials.

Composite Materials

Composite materials refer to a wide class of advanced materials which are composed of different materials or phases. The major idea is to obtain in total much better properties than a single component for itself could provide. Typical particularities include matrices (e.g., polymers, metals, and ceramic materials) which contain a second material in the form of particles or fibers as the reinforcing elements (see Fig. 1.1 for a schematic representation) or materials which are shaped in a particular way such as cellular materials (e.g., metal foams Ashby et al. (2000) or hollow sphere structures Öchsner and Augustin (2009)). In the case of cellular materials, the macroscopic properties are defined by the base material as well as the shape of the cells (voids, cavities, free space, and air) and their arrangement (e.g., periodic or stochastic).

For the particular group of fiber-reinforced materials, one may distinguish between classical fibers such as glass, carbon and boron Chawla (1987), natural fibers (e.g., flax, jute, and kenaf) Moudood et al. (2019), and nanofibers (e.g., carbon nanotubes) Yengejeh et al. (2017). The following chapters will focus on an important configuration, i.e., unidirectional fiber-reinforced thin layers. Such single layers or plies, also called laminae, provide superior mechanical properties compared to short fibers with random arrangement. The common approach is to stack several of such lamina with different orientations with respect to some reference direction to form a so-called laminate; see Fig. 1.2 for an example. Thus, the physical properties are dependent on the matrix material, the fibers, and the number/orientation/sequence (lay-up) of the plies which allows to tailor the macroscopic properties by adjusting the different

A. Öchsner and R. Makvandi, *A Numerical Approach to the Classical Laminate Theory of Composite Materials*, Advanced Structured Materials 189,
https://doi.org/10.1007/978-3-031-32975-3_1

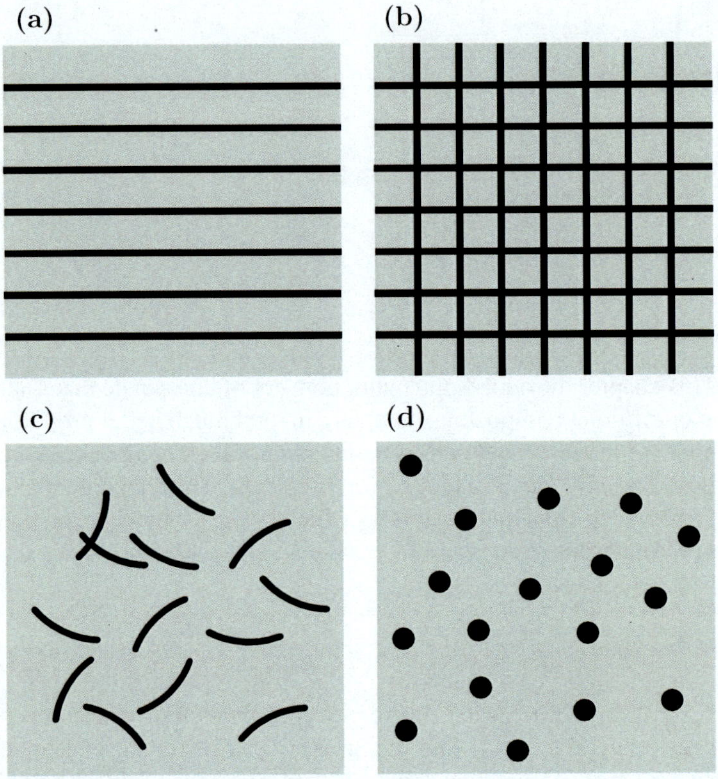

Fig. 1.1 Different types of composite materials (matrix = gray; reinforcement = black): **a** unidirectional fibers, **b** woven fibers, **c** short fibers, and **d** particles

parameters. This allows a much greater flexibility to adjust properties compared to classical engineering materials.

To clearly indicate the lay-up of a laminate, it is common to use the so-called laminate orientation code, i.e., some common rules to describe the different layers (laminae) of a laminate, see MIL-HDBK (2002) for further details:

- Square brackets are used to indicate the beginning and the end of the laminate code: [...].
- The orientation of each lamina is indicated by the angle between the fiber direction and the global x-axis.
- Orientations of successive laminae are separated by a virgule (/). They are listed in order from the first layer up to the last toward the positive z-direction: [.../.../...].
- Adjacent laminae with the same orientation are indicated by adding a subscript to the angle, equal to the number of repetitions.

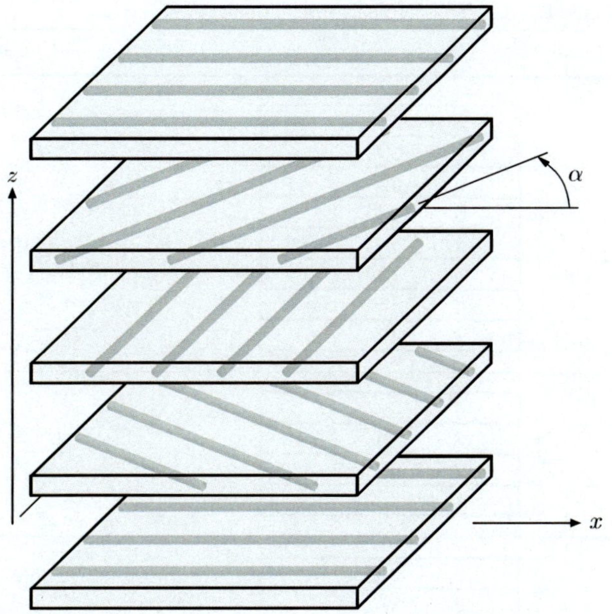

Fig. 1.2 Unbonded view of single laminae forming a 5-layer laminate

- A subscript of 's' is used if the first half of the lay-up is indicated and the second half is symmetric: $[\ldots/\ldots/\ldots]_{\text{S}}$. For a symmetric lay-up with an odd number of laminae, the layer which is not repeated is indicated by overlining the angle of that lamina: $[\ldots/\ldots/\overline{\ldots}]_{\text{S}}$.

These rules imply that each layer has the same material properties and thickness. Should this be not the case, more information must be provided. Some examples of different laminates and the corresponding laminate codes are provided in Table 1.1. It can be seen that the same laminate may be described by different representations, e.g., due to the symmetry or particular sequences.

Some of the textbooks which cover the mechanics of classical fiber-reinforced composite materials are summarized in Table 1.2. The interested reader may choose some of them as supplementary reading material to deepen and widen the knowledge presented in this textbook. Topics which are not covered in this textbook, e.g., manufacturing Hall and Javanbakht (2021) or the prediction of the properties of a single lamina (the so-called micromechanics) Herakovich (2012), Buryachenko (2007), and Dvorak (2013) can be found in the provided references at the end of this chapter.

Table 1.1 Example laminate orientation codes

Case	Cross section	Laminate code
(a)	z ↑ 8 0° 7 +60° 6 −60° 5 90° 4 90° 3 −45° 2 +45° $n=1$ 0°	$[0^\circ/45^\circ/-45^\circ/90^\circ/90^\circ/-60^\circ/60^\circ/0^\circ]$ $=[0^\circ/\pm 45^\circ/90^\circ_2/\mp 60^\circ/0^\circ]$
(b)	z ↑ 4 0° 3 90° 2 90° $n=1$ 0°	$[0^\circ/90^\circ/90^\circ/0^\circ]$ $=[0^\circ/90^\circ]_s$
(c)	z ↑ 3 90° 2 0° $n=1$ 90°	$[90^\circ/0^\circ/90^\circ]$ $=[90^\circ/\overline{0^\circ}]_s$

Table 1.2 Some of the English textbooks which cover the mechanics of classical fiber-reinforced composite materials (no claim to completeness)

Year (1st ed.)	Author	Title	References
1969	J. E. Ashton, J. C. Halpin, P. H. Petit	Primer on Composite Materials: Analysis	Ashton et al. (1969)
1969	L. R. Calcote	The Analysis of Laminated Composite Structures	Calcote (1969)
1970	J. E. Ashton, J. M. Whitney	Theory of Laminated Plates	Ashton and Whitney (1970)
1975	R. M. Jones	Mechanics of Composite Materials	Jones (1975)
1975	J. R. Vinson, T. W. Chou	Composite Materials and their Use in Structures	Vinson and Chou (1975)
1979	R. M. Christensen	Mechanics of Composite Materials	Christensen (1979)
1980	B. D. Agarwal, L. J. Broutman, K. Chandrashekhara	Analysis and Performance of Fiber Composites	Agarwal et al. (1980)
1980	S. W. Tsai, H. T. Hahn	Introduction to Composite Materials	Tsai and Hahn (1980)
1981	D. Hull	An Introduction to Composite Materials	Hull (1981)
1987	K. K. Chawla	Composite Materials: Science and Engineering	Chawla (1987)

(continued)

Table 1.2 (continued)

Year (1st ed.)	Author	Title	References
1987	J. R. Vinson, R. L. Sierakowski	The Behavior of Structures Composed of Composite Materials	Vinson and Sierakowski (1987)
1987	J. M. Whitney	Structural Analysis of Laminated Anisotropic Plates	Whitney (1980)
1990	V. K. S. Choo	Fundamentals of Composite Materials	Choo (1987)
1994	I. M. Daniel, O. Ishai	Engineering Mechanics of Composite Materials	Daniel and Ishai (1994)
1994	R. F. Gibson	Principles of Composite Material Mechanics	Gibson (1969)
1994	J. N. Reddy	Mechanics of Composite Materials: Selected Works of Nicholas J. Pagano	Reddy (1994)
1997	A. K. Kaw	Mechanics of Composite Materials	Kaw (1997)
1998	C. T. Herakovich	Mechanics of Fibrous Composites	Herakovich (1998)
1999	J.-M. Berthelot	Composite Materials: Mechanical Behavior and Structural Analysis	Berthelot (1999)
2001	V. V. Vasilev, E. V. Morozov	Mechanics and Analysis of Composite Materials	Vasiliev and Morozov (2001)
2002	C. Decolon	Analysis of Composite Structures	Decolon (2002)
2003	L. P. Kollár, G. S. Springer, W. Kissing	Mechanics of Composite Structures	Kollár and Springer (2003)
2004	H. Altenbach, J. Altenbach, W. Kissing	Mechanics of Composite Structural Elements	Altenbach et al. (2004)
2004	M. E. Tuttle	Structural Analysis of Polymeric Composite Materials	Tuttle (2004)
2008	R. de Borst, T. Sadowski	Lecture Notes on Composite Materials: Current Topics and Achievements	de Borst and Sadowski (2008)
2010	C. Kassapoglou	Design and Analysis of Composite Structures with Applications to Aerospace Structures	Kassapoglou (2010)
2021	A. Öchsner	Foundations of Classical Laminate Theory	Öchsner (2021)

References

Agarwal BD, Broutman LJ, Chandrashekhara K (1980) Analysis and performance of fiber composites. Wiley, New York

Altenbach H, Altenbach J, Kissing W (2004) Mechanics of composite structural elements. Springer, Berlin

Ashby M, Evans AG, Fleck NA, Gibson LJ, Hutchinson JW, Wadley HNG (2000) Metal foams: a design guide. Butterworth-Heinemann, Boston

Ashton JE, Halpin JC, Petit PH (1969) Primer on composite materials: analysis. Technomic Publishing Company, Stamford
Ashton JE, Whitney JM (1970) Theory of laminated plates. Technomic Publishing, Stamford
Berthelot J-M (1999) Composite materials: mechanical behavior and structural analysis. Springer, Berlin
Buryachenko VA (2007) Micromechanics of heterogeneous materials. Springer, New York
de Borst R, Sadowski T (2008) Lecture notes on composite materials: current topics and achievements. Springer, Dordrecht
Calcote LR (1969) The analysis of laminated composite structures. Van Nostrand Reinhold Company, New York
Chawla KK (1987) Composite materials: science and engineering. Springer, New York
Christensen RM (1979) Mechanics of composite materials. Wiley, New York
Choo VKS (1987) Fundamentals of composite materials. Knowen Academic Press, Delaware
Daniel IM, Ishai O (1994) Engineering mechanics of composite materials. Oxford University Press, New York
Decolon C (2002) Analysis of composite structures. Hermes Penton Science, London
Dvorak GJ (2013) Micromechanics of composite materials. Springer, New York
Gibson RF (1969) Principles of composite material mechanics. McGraw-Hill, New York
Hall W, Javanbakht Z (2021) Design and manufacture of fibre-reinforced composites. Springer, Cham
Herakovich CT (1998) Mechanics of fibrous composites. Wiley, New York
Herakovich CT (2012) Mechanics of composites: a historical review. Mech Res Commun 41:1–20
Hull D (1981) An introduction to composite materials. Cambridge University Press, Cambridge
Jones RM (1975) Mechanics of composite materials. Scripta Book Co, Washington
Kassapoglou C (2010) Design and analysis of composite structures with applications to aerospace structures. Wiley, Chichester
Kollár LP, Springer GS (2003) Mechanics of composite Structures. Cambridge University Press, Cambridge
Kaw AK (1997) Mechanics of composite materials. CRC Press, Boca Raton
Matthews FL, Davies GAO, Hitchings D, Soutis C (2000) Finite element modelling of composite materials and structures. Woodhead Publishing Limited, Cambridge
MIL-HDBK-17, 2F, (2002) Composite materials handbook-, vol 2. Polymer matrix composites materials properties, Department of Defence, USA
Moudood A, Rahman A, Khanlou HM, Hall W, Öchsner A, Francucci G (2019) Environmental effects on the durability and the mechanical performance of flax fiber/bio-epoxy composites. Compos Part B-Eng 171:284–293
Öchsner A, Augustin C (2009) Multifunctional metallic hollow sphere structures: manufacturing, properties and application. Springer, Berlin
Öchsner A (2021) Foundations of classical laminate theory. Springer, Cham
Reddy JN (1994) Mechanics of composite materials: selected works of Nicholas J. Kluwer Academic Publishers, Dordrecht, Pagano
Tsai SW, Hahn HT (1980) Introduction to composite materials. Technomic Publishishing Company, Lancaster
Tuttle ME (2004) Structural analysis of polymeric composite materials. Marcel Dekker, New York
Vasiliev VV, Morozov EV (2001) Mechanics and analysis of composite materials. Elsevier, Oxford
Vinson JR, Chou TW (1975) Composite materials and their use in structures. Wiley, New York
Vinson JR, Sierakowski RL (1987) The behavior of structures composed of composite materials. Kluwer Academic Publisher, Dordrecht
Whitney JM (1980) Structural analysis of laminated anisotropic plates. Technomic Publishing Company, Lancaster
Yengejeh SI, Kazemi SA, Öchsner A (2017) Carbon nanotubes as reinforcement in composites: a review of the analytical, numerical and experimental approaches. Comput Mater Sci 136:85–101

Chapter 2
Classical Laminate Theory

Abstract This chapter first covers the mechanical modeling of a single layer with unidirectionally aligned reinforcing fibers embedded in a homogeneous matrix, a so-called lamina. It is shown that a lamina can be treated as a combination of a plane elasticity element and a classical plate element. For both classical structural elements and their combination, the continuum mechanical modeling based on the three basic equations, i.e., the kinematics relationship, the constitutive law, and the equilibrium equation is presented. Combining these three questions results in the governing partial differential equations. The second part of the chapter covers the stacking of single laminae, generally under different angles, to a so-called laminate. Based on the approach of the classical laminate theory, a simplified stress analysis, and a subsequent failure analysis is derived, without the solution of the system of coupled differential equations for the unknown displacements in the three coordinate directions. This theory provides the solution of the statically indeterminate system based on a generalized stress-strain relationship under consideration of the constitutive relationship and the definition of the so-called stress resultants.

2.1 Continuum Mechanical Modeling

Typical thin laminates composed of orthotropic laminae (plies) as described in the previous section can be modeled under particular assumptions as a simple superposition of a plane elasticity and a classical plate element. These elements are composed of a combined plane elasticity and classical plate element, which is described in Sect. 2.2. Let us highlight here that the derivations in the following chapters follow a common approach from continuum mechanics Altenbach and Öchsner (2020); see Fig. 2.1.

A combination of the kinematics equation (i.e., the relation between the strains and deformations) with the constitutive equation (i.e., the relation between the stresses and strains) and the equilibrium equation (i.e., the equilibrium between the internal reactions and the external loads) results in a partial differential equation, or a corresponding system of differential equations. Limited to simple problems and config-

A. Öchsner and R. Makvandi, *A Numerical Approach to the Classical Laminate Theory of Composite Materials*, Advanced Structured Materials 189,
https://doi.org/10.1007/978-3-031-32975-3_2

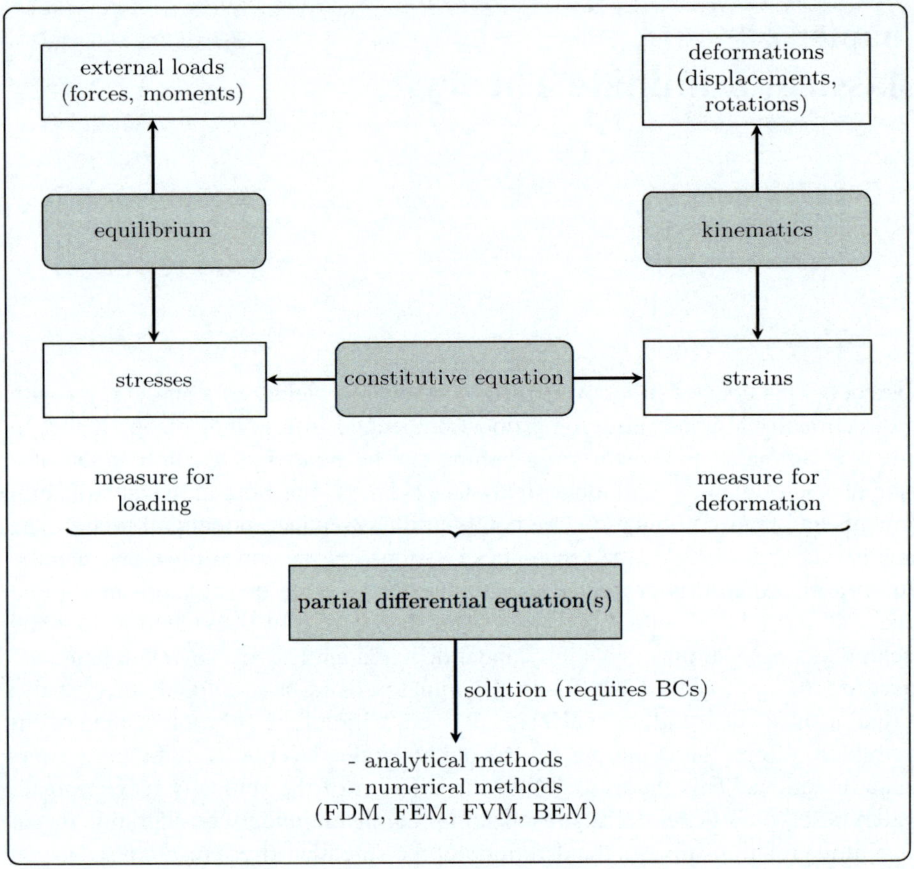

Fig. 2.1 Continuum mechanical modeling of structural members

urations, *analytical* solutions are possible. These analytical solutions are then exact in the frame of the assumptions made. However, the solution of complex problems requires the application of numerical methods such as the finite element method; see Öchsner (2020a), Öchsner and Öchsner (2018) for general details on the method and Table 2.1 for specialized literature on finite element simulations of composite materials. These *numerical* solutions are, in general, no longer exact since the numerical methods provide only an approximate solution. Thus, the major task of an engineer is then to ensure that the approximate solution is as good as possible. This requires a lot of experience and a solid foundation in the basics of applied mechanics, materials science, and mathematics.

The following sections present a simplified approach, the so-called classical laminate theory (CLT) Stravsky (1961), Dong et al. (1962), which aims to provide a stress and a subsequent strength/failure analysis without the solution of the system of coupled differential equations for the unknown displacements in the three coordinate directions. This theory provides the solution of the statically indeterminate system

Table 2.1 Some of the early textbooks which cover the finite element simulation of composite materials (no claim to completeness)

Year (1st ed.)	Author	Title	References
1992	O. O. OCHOA, J. N. REDDY	Finite Element Analysis of Composite Laminates	Ochoa and Reddy (1992)
1998	L. T. TENEK, J. ARGYRIS	Finite Element Analysis for Composite Structures	Tenek and Argyris (1998)
2000	F. L. MATTHEWS, G. A. O. DAVIES, D. HITCHINGS, C. SOUTIS	Finite Element Modelling of Composite Materials and Structures	Matthews et al. (2000)
2008	E. J. BARBERO	Finite Element Analysis of Composite Materials	Barbero (1996)

(i.e., no geometric boundary conditions in regard to u_1, u_2 and u_3 are considered) based on a generalized stress-strain relationship under consideration of the constitutive and kinematic relationships and the definition of the so-called stress resultants or generalized stresses and strains (i.e., the integral values of the corresponding stress distributions). Thus, external loads (either as distributed or concentrated loads) are not directly considered and must be balanced with the generalized stresses in order to be introduced.

The common assumptions of the classical laminate theory and the derivations in the following chapters can be summarized as follows (e.g., Daniel and Ishai 1994):

1. Each lamina is considered quasi-homogeneous and orthotropic (in general, the properties can range from isotropic to anisotropic).
2. Only unidirectional and flat lamina are considered in the following chapters.
3. The laminate consists of perfectly bonded laminae, and the bond lines are infinitesimally thin as well as non-shear-deformable.
4. The laminate is thin, i.e., the thickness is small compared to the lateral dimensions, and represents a state of plane stress.
5. Displacements (in thickness and lateral directions) are small compared to the thickness of the laminate.
6. Displacements are continuous throughout the laminate (non-shear-deformable bond lines).
7. In-plane displacements are linear functions of the thickness.
8. Shear strains in planes perpendicular to the middle surface are negligible.
9. Assumptions 7 and 8 imply that a line originally *straight* and *perpendicular* to the laminate middle surface remains so after deformation (Kirchhoff hypothesis of classical plate theory).
10. Kinematics and constitutive relations are linear.

11. Normal distances from the middle surface remain constant. Thus, the transverse normal strain is negligible compared to the in-plane normal strains (Kirchhoff hypothesis of classical plate theory).

2.2 Macromechanics of a Lamina

Let us consider in the following a unidirectional lamina as schematically shown in Fig. 2.2. The *1*-axis is aligned to the fibers, the *2*-axis is perpendicular to the fibers in the plane, and the *3*-axis indicates the thickness direction. Furthermore, the geometrical assumptions hold as outlined in Sect. 2.1, i.e., the lamina is assumed as flat and thin.

The general difference between a plane elasticity and a classical plate element in regard to the possible loads is illustrated in Fig. 2.3. A plane elasticity element can be only loaded by in-plane normal or shear forces; see Fig. 2.3a. Contrary to that, the classical plate element allows forces with lines of action in the thickness direction (*3*) or moments at which the moment vector lies in the *1*–*2* plane; see Fig. 2.3b. A similar distinction can be done in regard to the deformations: the plane elasticity element has only in-plane translations, whereas the classical plate element reveals translations perpendicular to the *1*–*2* plane and corresponding rotations. The derivations of the basic equations of continuum mechanics, i.e., the kinematics relationship, the constitutive law, and the equilibrium equation, are presented first separately for each of these members and afterwards superimposed to the combined plane elasticity and classical plate element. This approach allows for a better understanding of the continuum mechanical basics and the procedure to combine basic structural elements. Such a combination of structural elements is normally done in the early years of degree programs in mechanical engineering by superimposing the tensile rod with a thin beam to form the so-called generalized beam element which can elongate and bend Öchsner (2021).

The configuration shown in Fig. 2.3 is simply the two-dimensional generalization of the rod/beam superposition.

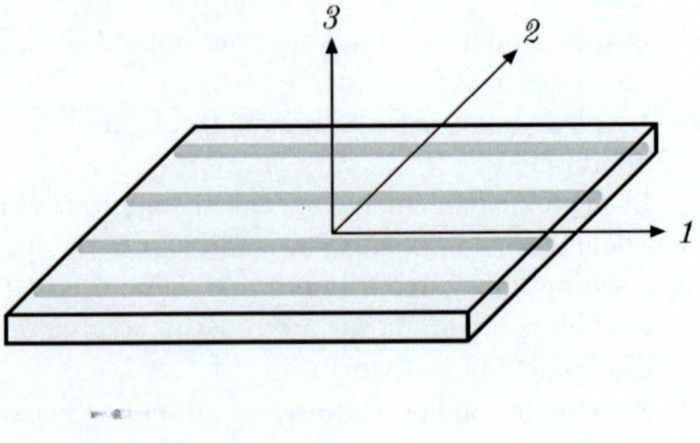

Fig. 2.2 Schematic representation of a unidirectional lamina

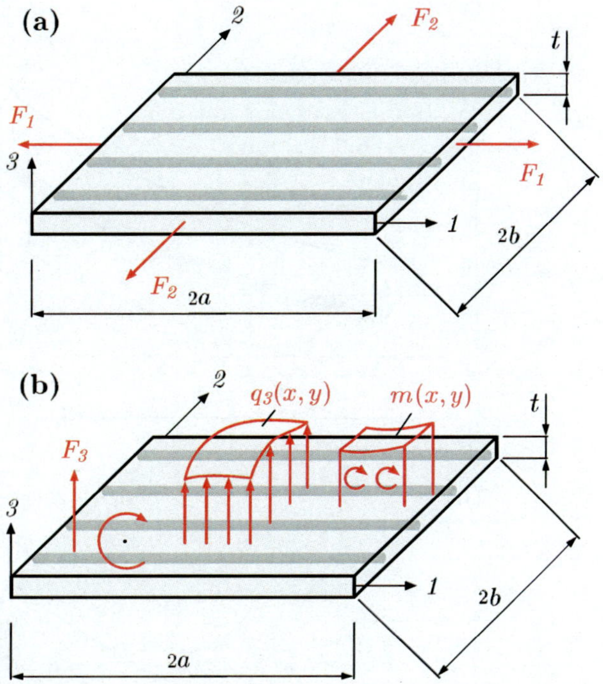

Fig. 2.3 General configuration for **a** a plane elasticity and **b** a classical plate element

2.2.1 Kinematics

The kinematics or strain-displacement relations extract the strain field contained in a displacement field.

Using the engineering definitions of strain, the following relations can be obtained for a plane elasticity element Chen and Han (1988), Eschenauer et al. (1997):

$$\varepsilon_1 = \frac{\partial u_1}{\partial 1}\,;\ \varepsilon_2 = \frac{\partial u_2}{\partial 2}\,;\ \gamma_{12} = 2\varepsilon_{12} = \frac{\partial u_1}{\partial 2} + \frac{\partial u_2}{\partial 1}\,. \tag{2.1}$$

In matrix notation, these three kinematics relationships can be written as

$$\begin{bmatrix} \varepsilon_1 \\ \varepsilon_2 \\ 2\varepsilon_{12} \end{bmatrix} = \begin{bmatrix} \frac{\partial}{\partial 1} & 0 \\ 0 & \frac{\partial}{\partial 2} \\ \frac{\partial}{\partial 2} & \frac{\partial}{\partial 1} \end{bmatrix} \begin{bmatrix} u_1 \\ u_2 \end{bmatrix}, \tag{2.2}$$

or symbolically as

$$\boldsymbol{\varepsilon} = \boldsymbol{\mathcal{L}}_1 \boldsymbol{u}_{1,2}\,, \tag{2.3}$$

where $\boldsymbol{\mathcal{L}}_1$ is the differential operator matrix.

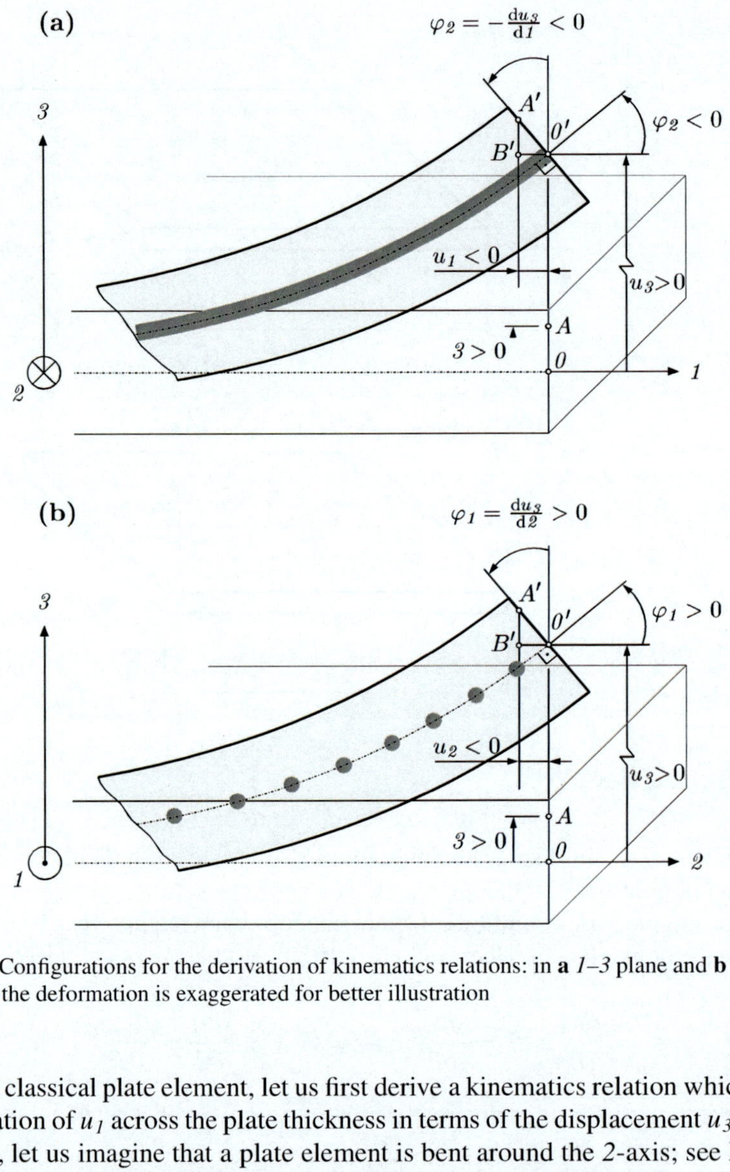

Fig. 2.4 Configurations for the derivation of kinematics relations: in **a** *1–3* plane and **b** *2–3* plane. Note that the deformation is exaggerated for better illustration

For a classical plate element, let us first derive a kinematics relation which relates the variation of u_1 across the plate thickness in terms of the displacement u_3. For this purpose, let us imagine that a plate element is bent around the *2*-axis; see Fig. 2.4a. The following definition of the rotational angle φ_2 is assumed: the angle φ_2 is positive if the vector of the rotational direction is pointing in positive *2*-axis.

Looking at the right-angled triangle $0'A'B'$, we can state that[1]

$$\sin(-\varphi_2) = \frac{\overline{B'0'}}{\overline{0'A'}} = \frac{-u_1}{3}, \tag{2.4}$$

[1] Note that according to the assumptions of the classical *thin* plate theory, the lengths $\overline{0A}$ and $\overline{0'A'}$ remain unchanged.

which results for small angles ($\sin(-\varphi_2) \approx -\varphi_2$) in

$$u_1 = +3\varphi_2\,. \tag{2.5}$$

Looking at the curved center line in Fig. 2.4a, it holds that the slope of the tangent line at $0'$ equals

$$\tan(-\varphi_2) = +\frac{\mathrm{d}u_3}{\mathrm{d}1} \approx -\varphi_2\,. \tag{2.6}$$

If Eqs. (2.5) and (2.6) are combined, it results as follows:

$$u_1 = -3\frac{\mathrm{d}u_3}{\mathrm{d}1}\,. \tag{2.7}$$

Considering a plate which is bent around the *1*-axis (see Fig. 2.4b) and following the same line of reasoning (the angle φ_1 is assumed positive if the vector of the rotational direction is pointing in positive *1*-axis.), similar equations can be derived for u_2:

$$\varphi_1 \approx \frac{\mathrm{d}u_3}{\mathrm{d}2}\,, \tag{2.8}$$

$$u_2 = -3\varphi_1\,, \tag{2.9}$$

$$u_2 = -3\frac{\mathrm{d}u_3}{\mathrm{d}2}\,. \tag{2.10}$$

One may find in the scholarly literature other definitions of the rotational angles Ventsel and Krauthammer (2001), Blaauwendraad (2010), Wang et al. (2000), Reddy (2006). The angle φ_2 is introduced in the *13*-plane (see Fig. 2.4a), whereas φ_1 is introduced in the *23*-plane (see Fig. 2.4b). These definitions are closer to the classical definitions of the angles in the scope of finite elements but do not conform with the definitions of the stress resultants (see M_1^{n} and M_2^{n} in Fig. 2.10). Other definitions assume, for example, that the rotational angle φ_1 (now defined in the *1–3* plane) is positive if it leads to a positive displacement u_1 at the positive *3*-side of the neutral axis. The same definition holds for the angle φ_2 (now defined in the *2–3* plane).

Using classical engineering definitions of strain, the following relations can be obtained Timoshenko and Woinowsky-Krieger (1959), Blaauwendraad (2010):

$$\varepsilon_1 = \frac{\partial u_1}{\partial 1} \overset{(2.7)}{=} \frac{\partial}{\partial 1}\left(-3\frac{\partial u_3}{\partial 1}\right) = -3\frac{\partial^2 u_3}{\partial 1^2} = 3\kappa_1 \,, \tag{2.11}$$

$$\varepsilon_2 = \frac{\partial u_2}{\partial 2} \overset{(2.10)}{=} \frac{\partial}{\partial 2}\left(-3\frac{\partial u_3}{\partial 2}\right) = -3\frac{\partial^2 u_3}{\partial 2^2} = 3\kappa_2 \,, \tag{2.12}$$

$$\gamma_{12} = \frac{\partial u_1}{\partial 2} + \frac{\partial u_2}{\partial 1} \overset{(2.7),(2.10)}{=} -23\frac{\partial^2 u_3}{\partial 1 \partial 2} = 3\kappa_{12} \,. \tag{2.13}$$

In matrix notation, these three relationships can be written as

$$\begin{bmatrix} \varepsilon_1 \\ \varepsilon_2 \\ \gamma_{12} \end{bmatrix} = -3 \begin{bmatrix} \frac{\partial^2}{\partial 1^2} \\ \frac{\partial^2}{\partial 2^2} \\ \frac{2\partial^2}{\partial 1 \partial 2} \end{bmatrix} u_3 = 3 \begin{bmatrix} \kappa_1 \\ \kappa_2 \\ \kappa_{12} \end{bmatrix} , \tag{2.14}$$

or symbolically as

$$\boldsymbol{\varepsilon} = -3\boldsymbol{\mathcal{L}}_2 u_3 = 3\boldsymbol{\kappa} \,. \tag{2.15}$$

Combining Eqs. (2.2) and (2.14), the following kinematics description for a superposed element is obtained as

$$\begin{bmatrix} \varepsilon_1 \\ \varepsilon_2 \\ \gamma_{12} \end{bmatrix} = \begin{bmatrix} \frac{\partial}{\partial 1} & 0 \\ 0 & \frac{\partial}{\partial 2} \\ \frac{\partial}{\partial 2} & \frac{\partial}{\partial 1} \end{bmatrix} \begin{bmatrix} u_1 \\ u_2 \end{bmatrix} - 3 \begin{bmatrix} \frac{\partial^2}{\partial 1^2} \\ \frac{\partial^2}{\partial 2^2} \\ \frac{2\partial^2}{\partial 1 \partial 2} \end{bmatrix} u_3 \tag{2.16}$$

$$= \begin{bmatrix} \frac{\partial}{\partial 1} & 0 \\ 0 & \frac{\partial}{\partial 2} \\ \frac{\partial}{\partial 2} & \frac{\partial}{\partial 1} \end{bmatrix} \begin{bmatrix} u_1 \\ u_2 \end{bmatrix} + 3 \begin{bmatrix} \kappa_1 \\ \kappa_2 \\ \kappa_{12} \end{bmatrix} , \tag{2.17}$$

or symbolically as

$$\boldsymbol{\varepsilon} = \boldsymbol{\mathcal{L}}_1 \boldsymbol{u}_{1,2} - 3\boldsymbol{\mathcal{L}}_2 u_3 \tag{2.18}$$

$$= \underbrace{\boldsymbol{\mathcal{L}}_1 \boldsymbol{u}_{1,2}}_{\boldsymbol{\varepsilon}^0} + 3\boldsymbol{\kappa} \,, \tag{2.19}$$

where $\boldsymbol{\varepsilon}^0$ collects the middle-surface strains. Alternatively, one may collect the single components of the plane elasticity and the plate contribution in matrix form as follows:

$$\begin{bmatrix} \varepsilon_1^0 \\ \varepsilon_2^0 \\ \gamma_{12}^0 \\ \hline \kappa_1 \\ \kappa_2 \\ \kappa_{12} \end{bmatrix} = \begin{bmatrix} \frac{\partial}{\partial 1} & 0 & 0 \\ 0 & \frac{\partial}{\partial 2} & 0 \\ \frac{\partial}{\partial 1} & \frac{\partial}{\partial 2} & 0 \\ \hline 0 & 0 & -\frac{\partial^2}{\partial 1^2} \\ 0 & 0 & -\frac{\partial^2}{\partial 2^2} \\ 0 & 0 & -\frac{2\partial^2}{\partial 1 \partial 2} \end{bmatrix} \begin{bmatrix} u_1 \\ u_2 \\ \hline u_3 \end{bmatrix}, \tag{2.20}$$

or symbolically as

$$\boldsymbol{e} = \boldsymbol{\mathcal{L}}'\boldsymbol{u}\,. \tag{2.21}$$

The column matrix of generalized strains can be also expressed as

$$\boldsymbol{e} = \begin{bmatrix} \boldsymbol{\varepsilon}^0 \\ \hline \boldsymbol{\kappa} \end{bmatrix}. \tag{2.22}$$

2.2.2 *Constitutive Equation*

For isotropic materials, the mechanical properties are the same for all directions, and it can be shown that the components of the elasticity matrix are determined by two independent constants. In mechanical engineering, many times the set of elasticity modulus E and Poisson's ratio ν are chosen as the independent constants. Orthotropic materials reveal three principal, mutually orthogonal axes. These axes are also called the material principal axes.

2.2.2.1 Isotropic Materials

For a *plane elasticity element*, the two-dimensional plane stress case ($\sigma_3 = \sigma_{23} = \sigma_{13} = 0$) shown in Fig. 2.5 is commonly used for the analysis of thin, flat plates loaded in the plane of the plate (*1*–*2* plane).

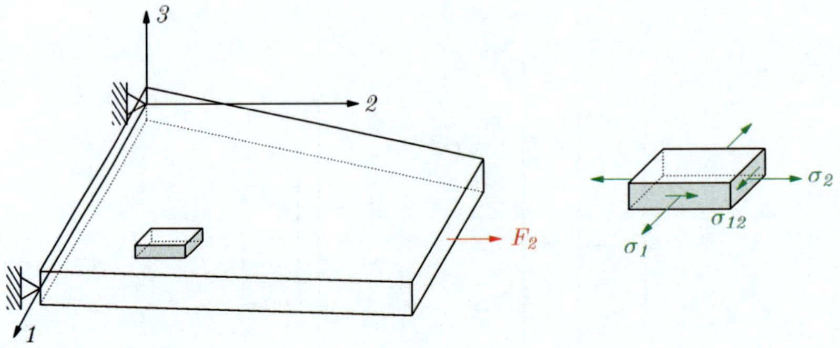

Fig. 2.5 Two-dimensional problem: plane stress case

It should be noted here that the normal thickness stress is zero ($\sigma_3 = 0$), whereas the thickness normal strain is present ($\varepsilon_3 \neq 0$).

The plane stress Hooke's law for a linear-elastic isotropic material based on the elasticity modulus E and Poisson's ratio ν can be written for constant temperature as

$$\begin{bmatrix} \sigma_1 \\ \sigma_2 \\ \sigma_{12} \end{bmatrix} = \frac{E}{1-\nu^2} \begin{bmatrix} 1 & \nu & 0 \\ \nu & 1 & 0 \\ 0 & 0 & \frac{1-\nu}{2} \end{bmatrix} \begin{bmatrix} \varepsilon_1 \\ \varepsilon_2 \\ 2\,\varepsilon_{12} \end{bmatrix}, \tag{2.23}$$

or in matrix notation as

$$\boldsymbol{\sigma} = \boldsymbol{C}\boldsymbol{\varepsilon}, \tag{2.24}$$

where $\boldsymbol{C}$ is the so-called elasticity matrix. It should be noted here that the engineering shear strain $\gamma_{12} = 2\varepsilon_{12}$ is used in the formulation of Eq. (2.23).

Rearranging the elastic stiffness form given in Eq. (2.23) for the strains gives the elastic compliance form

$$\begin{bmatrix} \varepsilon_1 \\ \varepsilon_2 \\ 2\,\varepsilon_{12} \end{bmatrix} = \frac{1}{E} \begin{bmatrix} 1 & -\nu & 0 \\ -\nu & 1 & 0 \\ 0 & 0 & 2(\nu+1) \end{bmatrix} \begin{bmatrix} \sigma_1 \\ \sigma_2 \\ \sigma_{12} \end{bmatrix}, \tag{2.25}$$

or in matrix notation as

$$\boldsymbol{\varepsilon} = \boldsymbol{D}\boldsymbol{\sigma}, \tag{2.26}$$

where $\boldsymbol{D} = \boldsymbol{C}^{-1}$ is the so-called elastic compliance matrix. The general characteristic of a plane Hooke's law in the form of Eqs. (2.24) and (2.25) is that two independent material parameters are used.

It should be finally noted that the thickness strain ε_3 can be obtained based on the two in-plane normal strains ε_1 and ε_2 as

$$\varepsilon_3 = -\frac{\nu}{1-\nu} \cdot \left(\varepsilon_1 + \varepsilon_2\right) . \tag{2.27}$$

The last equation can be derived from the three-dimensional formulation; see Öchsner (2020a). Equations (2.23) and (2.25) indicate that a plane isotropic material requires two independent material constants.[2]

The *classical plate theory* assumes a plane stress state, and the constitutive equation can be taken from Sect. 2.2.2.1, i.e., Eqs. (2.23) and (2.25) are still valid.

Since there is no difference in the constitutive description of a plane elasticity and a classical plate element, a combined element is still described based on the set of Eqs. (2.23)–(2.24) and (2.25)–(2.26).

2.2.2.2 Orthotropic Materials

Since the plane elasticity element follows the same constitutive description as the classical plate element, it is sufficient to indicate the relationship for the combined element. Let us start with the compliance form, i.e.

$$\begin{bmatrix} \varepsilon_1 \\ \varepsilon_2 \\ 2\,\varepsilon_{12} \end{bmatrix} = \begin{bmatrix} \frac{1}{E_1} & -\frac{\nu_{21}}{E_2} & 0 \\ -\frac{\nu_{12}}{E_1} & \frac{1}{E_2} & 0 \\ 0 & 0 & \frac{1}{G_{12}} \end{bmatrix} \begin{bmatrix} \sigma_1 \\ \sigma_2 \\ \sigma_{12} \end{bmatrix} = \begin{bmatrix} D_{11} & D_{12} & 0 \\ D_{12} & D_{22} & 0 \\ 0 & 0 & D_{44} \end{bmatrix} \begin{bmatrix} \sigma_1 \\ \sigma_2 \\ \sigma_{12} \end{bmatrix} , \tag{2.28}$$

or in matrix notation as

$$\boldsymbol{\varepsilon} = \boldsymbol{D}\boldsymbol{\sigma} , \tag{2.29}$$

where $\boldsymbol{D} = \boldsymbol{C}^{-1}$ is again the so-called elastic compliance matrix. The constant G in Eq. (2.29) is the shear modulus in the *1*–*2* plane. It should be noted that in Eq. (2.28), the identity

$$-\frac{\nu_{12}}{E_1} = -\frac{\nu_{21}}{E_2} \tag{2.30}$$

[2] Three-dimensional isotropy requires as well only two independent elastic constants.

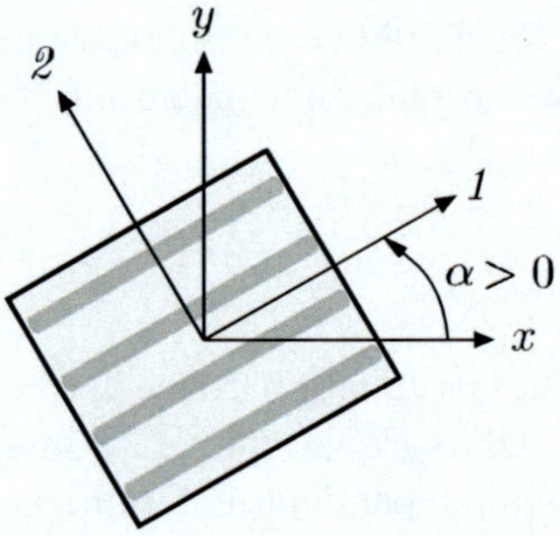

Fig. 2.6 Rotated lamina: (*1*, *2*) principal material coordinates and (x, y) arbitrary coordinates. The rotational angle α is from the x-axis to the *1*-axis (counterclockwise positive for the sketched coordinate systems)

holds, and we can conclude that four independent elastic constants are required to describe a plane orthotropic material.[3] The stress-strain relationship can be obtained from Eq. (2.28) by inverting the compliance matrix to give the following representation:

$$\begin{bmatrix} \sigma_1 \\ \sigma_2 \\ \sigma_{12} \end{bmatrix} = \begin{bmatrix} \dfrac{E_1}{1-\nu_{12}\nu_{21}} & \dfrac{\nu_{21}E_1}{1-\nu_{12}\nu_{21}} & 0 \\ \dfrac{\nu_{12}E_2}{1-\nu_{12}\nu_{21}} & \dfrac{E_2}{1-\nu_{12}\nu_{21}} & 0 \\ 0 & 0 & G_{12} \end{bmatrix} \begin{bmatrix} \varepsilon_1 \\ \varepsilon_2 \\ 2\varepsilon_{12} \end{bmatrix} = \begin{bmatrix} C_{11} & C_{12} & 0 \\ C_{12} & C_{22} & 0 \\ 0 & 0 & C_{44} \end{bmatrix} \begin{bmatrix} \varepsilon_1 \\ \varepsilon_2 \\ 2\varepsilon_{12} \end{bmatrix}, \tag{2.31}$$

or in matrix notation as

$$\boldsymbol{\sigma} = \boldsymbol{C}\boldsymbol{\varepsilon}, \tag{2.32}$$

where $\boldsymbol{C} = \boldsymbol{D}^{-1}$ is again the so-called elasticity matrix.

Let us consider in the following the rotation of the coordinate system and the corresponding transformations of stresses and strains; see Fig. 2.6. This is important in the case of laminates with different laminae at different orientations (see Sect. 2.3):

$$\begin{bmatrix} \sigma_x \\ \sigma_y \\ \sigma_{xy} \end{bmatrix} = \begin{bmatrix} \cos^2\alpha & \sin^2\alpha & -2\sin\alpha\cos\alpha \\ \sin^2\alpha & \cos^2\alpha & 2\sin\alpha\cos\alpha \\ \sin\alpha\cos\alpha & -\sin\alpha\cos\alpha & \cos^2\alpha - \sin^2\alpha \end{bmatrix} \begin{bmatrix} \sigma_1 \\ \sigma_2 \\ \sigma_{12} \end{bmatrix}, \tag{2.33}$$

or in matrix notation as

$$\boldsymbol{\sigma}_{x,y} = \boldsymbol{T}_{\sigma}^{-1}\boldsymbol{\sigma}_{1,2}, \tag{2.34}$$

or in the inverse representation

[3] Three-dimensional orthotropy requires nine independent elastic constants.

$$\begin{bmatrix} \sigma_1 \\ \sigma_2 \\ \sigma_{12} \end{bmatrix} = \begin{bmatrix} \cos^2\alpha & \sin^2\alpha & 2\sin\alpha\cos\alpha \\ \sin^2\alpha & \cos^2\alpha & -2\sin\alpha\cos\alpha \\ -\sin\alpha\cos\alpha & \sin\alpha\cos\alpha & \cos^2\alpha - \sin^2\alpha \end{bmatrix} \begin{bmatrix} \sigma_x \\ \sigma_y \\ \sigma_{xy} \end{bmatrix}, \tag{2.35}$$

or in matrix notation as

$$\boldsymbol{\sigma}_{1,2} = \boldsymbol{T}_{\sigma}\boldsymbol{\sigma}_{x,y}\,. \tag{2.36}$$

The corresponding transformations for the strain field read

$$\begin{bmatrix} \varepsilon_x \\ \varepsilon_y \\ 2\varepsilon_{xy} \end{bmatrix} = \begin{bmatrix} \cos^2\alpha & \sin^2\alpha & -\sin\alpha\cos\alpha \\ \sin^2\alpha & \cos^2\alpha & \sin\alpha\cos\alpha \\ 2\sin\alpha\cos\alpha & -2\sin\alpha\cos\alpha & \cos^2\alpha - \sin^2\alpha \end{bmatrix} \begin{bmatrix} \varepsilon_1 \\ \varepsilon_2 \\ 2\varepsilon_{12} \end{bmatrix}, \tag{2.37}$$

or in matrix notation as

$$\boldsymbol{\varepsilon}_{x,y} = \boldsymbol{T}_{\varepsilon}^{-1}\boldsymbol{\varepsilon}_{1,2}\,, \tag{2.38}$$

or in the inverse representation

$$\begin{bmatrix} \varepsilon_1 \\ \varepsilon_2 \\ 2\varepsilon_{12} \end{bmatrix} = \begin{bmatrix} \cos^2\alpha & \sin^2\alpha & \sin\alpha\cos\alpha \\ \sin^2\alpha & \cos^2\alpha & -\sin\alpha\cos\alpha \\ -2\sin\alpha\cos\alpha & 2\sin\alpha\cos\alpha & \cos^2\alpha - \sin^2\alpha \end{bmatrix} \begin{bmatrix} \varepsilon_x \\ \varepsilon_y \\ 2\varepsilon_{xy} \end{bmatrix}, \tag{2.39}$$

or in matrix notation as

$$\boldsymbol{\varepsilon}_{1,2} = \boldsymbol{T}_{\varepsilon}\boldsymbol{\varepsilon}_{x,y}\,. \tag{2.40}$$

It should be noted here that the relationship between the strain and stress transformation matrices is as follows:

$$\boldsymbol{T}_{\varepsilon} = \boldsymbol{R}\boldsymbol{T}_{\sigma}\boldsymbol{R}^{-1}\,, \tag{2.41}$$

where the matrix $\boldsymbol{R}$ has the following entries:

$$\boldsymbol{R} = \begin{bmatrix} 1 & 0 & 0 \\ 0 & 1 & 0 \\ 0 & 0 & 2 \end{bmatrix}. \tag{2.42}$$

Based on these transformations of the stresses and strains, the stress-strain relationship in the arbitrary (rotated) x–y coordinate system can be obtained as

$$\boldsymbol{\sigma}_{1,2} = \boldsymbol{C}\boldsymbol{\varepsilon}_{1,2}\,, \tag{2.43}$$

$$\underbrace{\boldsymbol{T}_{\sigma}^{-1}\boldsymbol{\sigma}_{1,2}}_{\boldsymbol{\sigma}_{x,y}} = \boldsymbol{T}_{\sigma}^{-1}\boldsymbol{C}\boldsymbol{\varepsilon}_{1,2}\,, \tag{2.44}$$

$$\boldsymbol{\sigma}_{x,y} = \boldsymbol{T}_{\sigma}^{-1}\boldsymbol{C}\underbrace{\boldsymbol{T}_{\varepsilon}\boldsymbol{T}_{\varepsilon}^{-1}}_{\mathbf{I}}\boldsymbol{\varepsilon}_{1,2}\,, \tag{2.45}$$

$$= \boldsymbol{T}_{\sigma}^{-1}\boldsymbol{C}\boldsymbol{T}_{\varepsilon}\underbrace{\boldsymbol{T}_{\varepsilon}^{-1}\boldsymbol{\varepsilon}_{1,2}}_{\boldsymbol{\varepsilon}_{x,y}}\,, \tag{2.46}$$

$$\boldsymbol{\sigma}_{x,y} = \underbrace{\boldsymbol{T}_{\sigma}^{-1}\boldsymbol{C}\boldsymbol{T}_{\varepsilon}}_{\overline{\boldsymbol{C}}}\boldsymbol{\varepsilon}_{x,y}\,. \tag{2.47}$$

Thus, the transformed elasticity matrix $\overline{\boldsymbol{C}}$ can be obtained from the following triple matrix product:

$$\overline{\boldsymbol{C}} = \begin{bmatrix} \cos^2\alpha & \sin^2\alpha & -2\sin\alpha\cos\alpha \\ \sin^2\alpha & \cos^2\alpha & 2\sin\alpha\cos\alpha \\ \sin\alpha\cos\alpha & -\sin\alpha\cos\alpha & \cos^2\alpha-\sin^2\alpha \end{bmatrix} \begin{bmatrix} C_{11} & C_{12} & C_{14} \\ C_{12} & C_{22} & C_{24} \\ C_{14} & C_{24} & C_{44} \end{bmatrix}$$

$$\begin{bmatrix} \cos^2\alpha & \sin^2\alpha & \sin\alpha\cos\alpha \\ \sin^2\alpha & \cos^2\alpha & -\sin\alpha\cos\alpha \\ -2\sin\alpha\cos\alpha & 2\sin\alpha\cos\alpha & \cos^2\alpha-\sin^2\alpha \end{bmatrix} \tag{2.48}$$

$$= \begin{bmatrix} \cos^2\alpha & \sin^2\alpha & -2\sin\alpha\cos\alpha \\ \sin^2\alpha & \cos^2\alpha & 2\sin\alpha\cos\alpha \\ \sin\alpha\cos\alpha & -\sin\alpha\cos\alpha & \cos^2\alpha-\sin^2\alpha \end{bmatrix} \begin{bmatrix} \frac{E_1}{1-\nu_{12}\nu_{21}} & \frac{\nu_{21}E_1}{1-\nu_{12}\nu_{21}} & 0 \\ \frac{\nu_{12}E_2}{1-\nu_{12}\nu_{21}} & \frac{E_2}{1-\nu_{12}\nu_{21}} & 0 \\ 0 & 0 & G_{12} \end{bmatrix}$$

$$\begin{bmatrix} \cos^2\alpha & \sin^2\alpha & \sin\alpha\cos\alpha \\ \sin^2\alpha & \cos^2\alpha & -\sin\alpha\cos\alpha \\ -2\sin\alpha\cos\alpha & 2\sin\alpha\cos\alpha & \cos^2\alpha-\sin^2\alpha \end{bmatrix} \tag{2.49}$$

$$= \begin{bmatrix} \overline{C}_{11} & \overline{C}_{12} & \overline{C}_{14} \\ \overline{C}_{12} & \overline{C}_{22} & \overline{C}_{24} \\ \overline{C}_{14} & \overline{C}_{24} & \overline{C}_{44} \end{bmatrix}, \tag{2.50}$$

where the single matrix entries $\overline{C}_{ij}$ are given as follows:

$$\overline{C}_{11} = C_{11}\cos^4\alpha + 2(C_{12} + 2C_{44})\sin^2\alpha\cos^2\alpha + C_{22}\sin^4\alpha\,, \tag{2.51}$$

$$\overline{C}_{12} = (C_{11} + C_{22} - 4C_{44})\sin^2\alpha\cos^2 + C_{12}(\sin^4\alpha + \cos^4\alpha)\,, \tag{2.52}$$

$$\overline{C}_{22} = C_{11}\sin^4\alpha + 2(C_{12} + 2C_{44})\sin^2\alpha\cos^2\alpha + C_{22}\cos^4\alpha\,, \tag{2.53}$$

$$\overline{C}_{14} = (C_{11} - C_{12} - 2C_{44})\sin\alpha\cos^3\alpha + (C_{12} - C_{22} + 2C_{44})\sin^3\alpha\cos\alpha\,, \tag{2.54}$$

$$\overline{C}_{24} = (C_{11} - C_{12} - 2C_{44})\sin^3\alpha\cos\alpha + (C_{12} - C_{22} + 2C_{44})\sin\alpha\cos^3\alpha\,, \tag{2.55}$$

$$\overline{C}_{44} = (C_{11} + C_{22} - 2C_{12} - 2C_{44})\sin^2\alpha\cos^2\alpha + C_{44}(\sin^4\alpha + \cos^4\alpha)\,. \tag{2.56}$$

In a similar way, the strain-stress relationship in the x–y coordinate system can be obtained as

$$\boldsymbol{\varepsilon}_{1,2} = \boldsymbol{D}\boldsymbol{\sigma}_{1,2}\,, \tag{2.57}$$

$$\underbrace{\boldsymbol{T}_{\varepsilon}^{-1}\boldsymbol{\varepsilon}_{1,2}}_{\boldsymbol{\varepsilon}_{x,y}} = \boldsymbol{T}_{\varepsilon}^{-1}\boldsymbol{D}\boldsymbol{\sigma}_{1,2}\,, \tag{2.58}$$

$$\boldsymbol{\varepsilon}_{x,y} = \boldsymbol{T}_{\varepsilon}^{-1}\boldsymbol{D}\underbrace{\boldsymbol{T}_{\sigma}\boldsymbol{T}_{\sigma}^{-1}}_{\mathbf{I}}\boldsymbol{\sigma}_{1,2}\,, \tag{2.59}$$

$$= \boldsymbol{T}_{\epsilon}^{-1}\boldsymbol{D}\boldsymbol{T}_{\sigma}\underbrace{\boldsymbol{T}_{\sigma}^{-1}\boldsymbol{\sigma}_{1,2}}_{\boldsymbol{\sigma}_{x,y}}\,, \tag{2.60}$$

$$\boldsymbol{\varepsilon}_{x,y} = \underbrace{\boldsymbol{T}_{\varepsilon}^{-1}\boldsymbol{D}\boldsymbol{T}_{\sigma}}_{\overline{\boldsymbol{D}}}\boldsymbol{\sigma}_{x,y}\,. \tag{2.61}$$

Thus, the transformed elastic compliance matrix $\overline{\boldsymbol{D}}$ can be obtained from the following triple matrix product:

$$\overline{\boldsymbol{D}} = \begin{bmatrix} \cos^2\alpha & \sin^2\alpha & -\sin\alpha\cos\alpha \\ \sin^2\alpha & \cos^2\alpha & \sin\alpha\cos\alpha \\ 2\sin\alpha\cos\alpha & -2\sin\alpha\cos\alpha & \cos^2\alpha - \sin^2\alpha \end{bmatrix} \begin{bmatrix} D_{11} & D_{12} & D_{14} \\ D_{12} & D_{22} & D_{24} \\ D_{14} & D_{24} & D_{44} \end{bmatrix}$$
$$\begin{bmatrix} \cos^2\alpha & \sin^2\alpha & 2\sin\alpha\cos\alpha \\ \sin^2\alpha & \cos^2\alpha & -2\sin\alpha\cos\alpha \\ -\sin\alpha\cos\alpha & \sin\alpha\cos\alpha & \cos^2\alpha - \sin^2\alpha \end{bmatrix} \tag{2.62}$$

$$= \begin{bmatrix} \cos^2\alpha & \sin^2\alpha & -\sin\alpha\cos\alpha \\ \sin^2\alpha & \cos^2\alpha & \sin\alpha\cos\alpha \\ 2\sin\alpha\cos\alpha & -2\sin\alpha\cos\alpha & \cos^2\alpha - \sin^2\alpha \end{bmatrix} \begin{bmatrix} \frac{1}{E_1} & -\frac{\nu_{21}}{E_2} & 0 \\ -\frac{\nu_{12}}{E_1} & \frac{1}{E_2} & 0 \\ 0 & 0 & \frac{1}{G_{12}} \end{bmatrix}$$

$$\begin{bmatrix} \cos^2\alpha & \sin^2\alpha & 2\sin\alpha\cos\alpha \\ \sin^2\alpha & \cos^2\alpha & -2\sin\alpha\cos\alpha \\ -\sin\alpha\cos\alpha & \sin\alpha\cos\alpha & \cos^2\alpha - \sin^2\alpha \end{bmatrix} \tag{2.63}$$

$$= \begin{bmatrix} \overline{D}_{11} & \overline{D}_{12} & \overline{D}_{14} \\ \overline{D}_{12} & \overline{D}_{22} & \overline{D}_{24} \\ \overline{D}_{14} & \overline{D}_{24} & \overline{D}_{44} \end{bmatrix}, \tag{2.64}$$

where the single matrix entries $\overline{D}_{ij}$ are given as follows:

$$\overline{D}_{11} = D_{11}\cos^4\alpha + (2D_{12} + D_{44})\sin^2\alpha\cos^2\alpha + D_{22}\sin^4\alpha\,, \tag{2.65}$$

$$\overline{D}_{12} = (D_{11} + D_{22} - D_{44})\sin^2\alpha\cos^2 + D_{12}(\sin^4\alpha + \cos^4\alpha)\,, \tag{2.66}$$

$$\overline{D}_{22} = D_{11}\sin^4\alpha + (2D_{12} + D_{44})\sin^2\alpha\cos^2\alpha + D_{22}\cos^4\alpha\,, \tag{2.67}$$

$$\overline{D}_{14} = (2D_{11} - 2D_{12} - D_{44})\sin\alpha\cos^3\alpha - (2D_{22} - 2D_{12} - D_{44})\sin^3\alpha\cos\alpha\,, \tag{2.68}$$

$$\overline{D}_{24} = (2D_{11} - 2D_{12} - D_{44})\sin^3\alpha\cos\alpha - (2D_{22} - 2D_{12} - D_{44})\sin\alpha\cos^3\alpha\,, \tag{2.69}$$

$$\overline{D}_{44} = 2(2D_{11} + 2D_{22} - 4D_{12} - D_{44})\sin^2\alpha\cos^2\alpha + D_{44}(\sin^4\alpha + \cos^4\alpha)\,. \tag{2.70}$$

It can be shown that the elements $\overline{D}_{ij}$ of the transformed elastic compliance matrix (see Eq. (2.64)) can be related to the apparent engineering constants of the orthotropic lamina in the rotated x–y coordinate system as follows (see Jones (1975), Daniel and Ishai (1994), Ashton and Whitney (1970), and Clyne and Hull (2019) for details):

$$\overline{\boldsymbol{D}} = \begin{bmatrix} \overline{D}_{11} & \overline{D}_{12} & \overline{D}_{14} \\ \overline{D}_{12} & \overline{D}_{22} & \overline{D}_{24} \\ \overline{D}_{14} & \overline{D}_{24} & \overline{D}_{44} \end{bmatrix} = \begin{bmatrix} \frac{1}{E_x} & -\frac{\nu_{xy}}{E_x} & \frac{\eta_x}{E_x} \\ -\frac{\nu_{xy}}{E_x} & \frac{1}{E_y} & \frac{\eta_y}{E_y} \\ \frac{\eta_x}{E_x} & \frac{\eta_y}{E_y} & \frac{1}{G_{xy}} \end{bmatrix}, \tag{2.71}$$

where the shear-extension coupling coefficients are generally defined as follows:

$$\eta_x = \frac{\gamma_{xy}}{\varepsilon_x} = \frac{\gamma_{xy}E_x}{\sigma_x}, \quad \text{(for uniaxial tension with } \sigma_x\text{)} \tag{2.72}$$

$$\eta_y = \frac{\gamma_{xy}}{\varepsilon_y} = \frac{\gamma_{xy}E_y}{\sigma_y}, \quad \text{(for uniaxial tension with } \sigma_y\text{).} \tag{2.73}$$

Furthermore, Poisson's ratios (extension-extension coupling coefficients) are defined as

$$\nu_{xy} = -\frac{\varepsilon_y}{\varepsilon_x}, \quad \text{(for uniaxial tension with } \sigma_x) \tag{2.74}$$

$$\nu_{yx} = -\frac{\varepsilon_x}{\varepsilon_y}. \quad \text{(for uniaxial tension with } \sigma_y). \tag{2.75}$$

2.2.3 *Equilibrium*

The equilibrium equations relate the external loads (i.e., forces and moments) to the corresponding internal reactions (i.e., stresses).

For a *plane elasticity element*, Fig. 2.7 shows the normal and shear stresses which are acting on a differential volume element in the *1*-direction. All forces are drawn in their positive direction at each cut face. A positive cut face is obtained if the outward surface normal is directed in the positive direction of the corresponding coordinate axis. This means that the right-hand face in Fig. 2.7 is positive and the force $(\sigma_1 + \frac{\partial \sigma_1}{\partial 1}\mathrm{d}1)\mathrm{d}2\mathrm{d}3$ is oriented in the positive *1*-direction. In a similar way, the top face is positive, i.e., the outward surface normal is directed in the positive *2*-direction, and the shear force[4] is oriented in the positive *1*-direction. Since the volume element is assumed to be in equilibrium, forces resulting from stresses on the sides of the cuboid and from the body forces f_i $(i = 1, 2, 3)$ must be balanced. These body forces are defined as forces per unit volume which can be produced by gravity,[5] acceleration, magnetic fields, and so on.

The static equilibrium of forces in the *1*-direction based on the five force components—two normal forces, two shear forces, and one body force—indicated in Fig. 2.7 gives

$$\left(\sigma_1 + \frac{\partial \sigma_1}{\partial 1}\right)\mathrm{d}2\mathrm{d}3 - \sigma_1\mathrm{d}2\mathrm{d}3 + \left(\frac{\partial \sigma_{21}}{\partial 2}\right)\mathrm{d}1\mathrm{d}3$$
$$- \sigma_{21}\mathrm{d}1\mathrm{d}3 + f_1\mathrm{d}1\mathrm{d}2\mathrm{d}3 = 0\,, \tag{2.76}$$

or after simplification and canceling with $\mathrm{d}V = \mathrm{d}1\mathrm{d}2\mathrm{d}3$:

$$\frac{\partial \sigma_1}{\partial 1} + \frac{\partial \sigma_{21}}{\partial 2} + f_1 = 0\,. \tag{2.77}$$

---

[4] In the case of a shear force σ_{ij}, the first index i indicates that the stress acts on a plane normal to the i-axis and the second index j denotes the direction in which the stress acts.

[5] If gravity is acting, the body force f results as the product of density times standard gravity: $f = \frac{F}{V} = \frac{mg}{V} = \frac{m}{V}g = \varrho g$. The units can be checked by consideration of $1\,\mathrm{N} = 1\frac{\mathrm{mkg}}{\mathrm{s}^2}$.

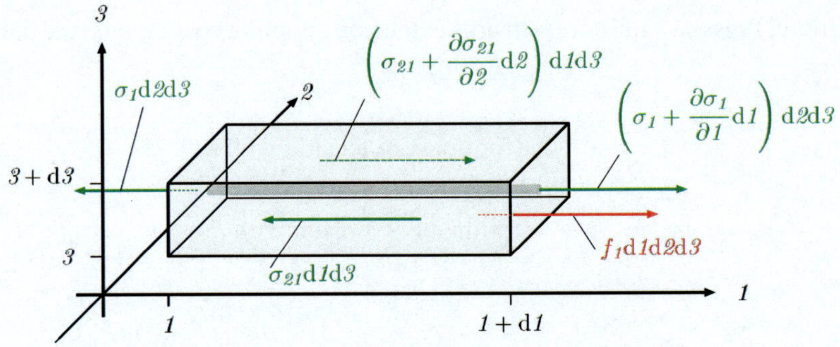

Fig. 2.7 Stresses and body forces which act on a plane differential volume element in *1*-direction (note that the three directions d*1*, d*2*, and d*3* are differently sketched to indicate the plane problem)

Based on the same approach, a similar equation can be specified in the *2*-direction:

$$\frac{\partial \sigma_2}{\partial 2} + \frac{\partial \sigma_{21}}{\partial 1} + f_2 = 0\,. \tag{2.78}$$

These two balance equations can be written in matrix notation as

$$\begin{bmatrix} \frac{\partial}{\partial 1} & 0 & \frac{\partial}{\partial 2} \\ 0 & \frac{\partial}{\partial 2} & \frac{\partial}{\partial 1} \end{bmatrix} \begin{bmatrix} \sigma_1 \\ \sigma_2 \\ \sigma_{12} \end{bmatrix} + \begin{bmatrix} f_1 \\ f_2 \end{bmatrix} = \begin{bmatrix} 0 \\ 0 \end{bmatrix}, \tag{2.79}$$

or in symbolic notation:

$$\boldsymbol{\mathcal{L}}_1^{\mathrm{T}} \boldsymbol{\sigma} + \boldsymbol{b} = \boldsymbol{0}\,, \tag{2.80}$$

where $\boldsymbol{\mathcal{L}}_1$ is the differential operator matrix and $\boldsymbol{b}$ is the column matrix of body forces.

The internal reactions expressed as stresses as indicated in Fig. 2.7 can be alternatively stated as forces based on an integration over the corresponding reference area. These integral values are then called the stress resultants or generalized stresses.

Based on Fig. 2.8, the stress resultant N_i^{n}, i.e., normalized with the corresponding side length of the plate element, for the plane elasticity element can be indicated as follows:

$$N_1^{\mathrm{n}} = \frac{N_1}{\mathrm{d}2} = \int_{-t/2}^{t/2} \sigma_1 \mathrm{d}3\,, \tag{2.81}$$

$$N_2^{\mathrm{n}} = \frac{N_2}{\mathrm{d}1} = \int_{-t/2}^{t/2} \sigma_2 \mathrm{d}3\,, \tag{2.82}$$

Fig. 2.8 Stress resultants for a plane elasticity element

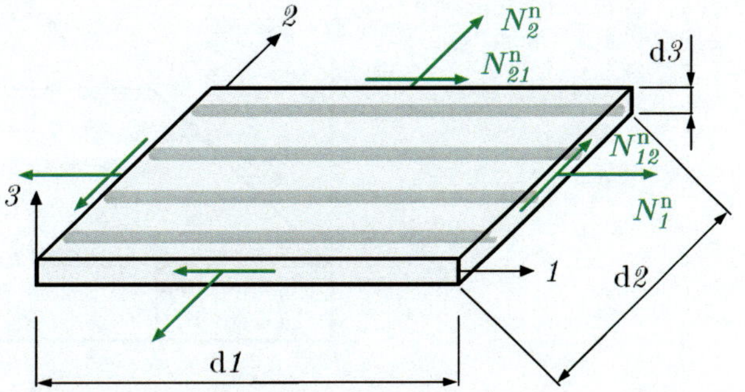

$$N^{\mathrm{n}}_{12} = \frac{N_{12}}{\mathrm{d}1} = \int_{-t/2}^{t/2} \sigma_{12}\mathrm{d}3\,, \tag{2.83}$$

or all three relations combined in matrix notation:

$$\begin{bmatrix} N^{\mathrm{n}}_{1} \\ N^{\mathrm{n}}_{2} \\ N^{\mathrm{n}}_{12} \end{bmatrix} = \int_{-t/2}^{t/2} \begin{bmatrix} \sigma_{1} \\ \sigma_{2} \\ \sigma_{12} \end{bmatrix} \mathrm{d}3\,. \tag{2.84}$$

Thus, we can get the following alternative formulation of the balance equations by introducing the stress resultants in Eq. (2.79). The two balance equations can be alternatively written in matrix notation as

$$\begin{bmatrix} \frac{\partial}{\partial 1} & 0 & \frac{\partial}{\partial 2} \\ 0 & \frac{\partial}{\partial 2} & \frac{\partial}{\partial 1} \end{bmatrix} \begin{bmatrix} N^{\mathrm{n}}_{1} \\ N^{\mathrm{n}}_{2} \\ N^{\mathrm{n}}_{12} \end{bmatrix} + \begin{bmatrix} q_{1} \\ q_{2} \end{bmatrix} = \begin{bmatrix} 0 \\ 0 \end{bmatrix}, \tag{2.85}$$

or in symbolic notation:

$$\boldsymbol{\mathcal{L}}_1^{\mathrm{T}} \boldsymbol{N}^{\mathrm{n}} + \boldsymbol{q}_{1,2} = \boldsymbol{0}\,, \tag{2.86}$$

where $q_1 = f_1\mathrm{d}3$ and $q_2 = f_2\mathrm{d}3$ are the area-specific loads.

For a *classical plate element*, let us first look at the stress distributions through the thickness of a classical plate element $\mathrm{d}1\mathrm{d}2t$ as shown in Fig. 2.9. Linear distributed normal stresses (σ_1, σ_2), linear distributed shear stresses (τ_{21}, τ_{12}), and parabolic distributed shear stresses (τ_{23}, τ_{13}) can be identified. These stresses can be expressed by the so-called stress resultants, i.e., bending moments and shear forces as shown in Fig. 2.10. These stress resultants are taken to be positive if they cause a tensile stress (positive) at a point with positive 3-coordinate.

These stress resultants are obtained by integrating over the stress distributions. In the case of plates, however, the integration is only performed over the thickness, i.e.,

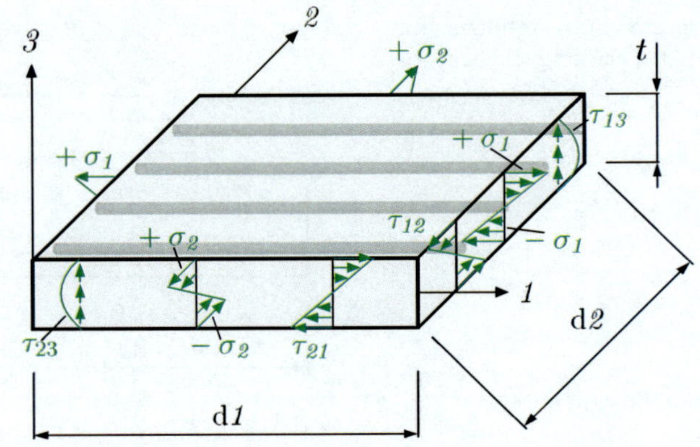

Fig. 2.9 Stresses acting on a classical plate element

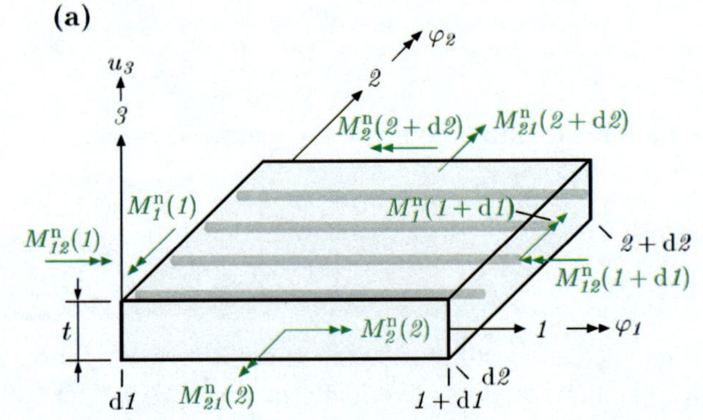

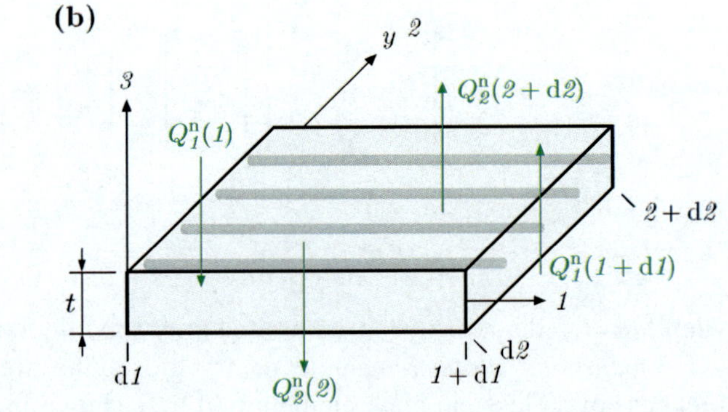

Fig. 2.10 Stress resultants acting on a classical plate element: **(a)** bending and twisting moments and **(b)** shear forces. Positive directions are drawn

the moments and forces are given per unit length (normalized with the corresponding side length of the plate element).

The normalized (superscript 'n') bending moments are obtained as

$$M_1^{\mathrm{n}} = \frac{M_1}{\mathrm{d}2} = \int_{-t/2}^{t/2} 3\sigma_1 \mathrm{d}3\,, \tag{2.87}$$

$$M_2^{\mathrm{n}} = \frac{M_2}{\mathrm{d}1} = \int_{-t/2}^{t/2} 3\sigma_2 \mathrm{d}3\,. \tag{2.88}$$

The twisting moment per unit length reads

$$M_{12}^{\mathrm{n}} = M_{21}^{\mathrm{n}} = \frac{M_{12}}{\mathrm{d}2} = \frac{M_{21}}{\mathrm{d}1} = \int_{-t/2}^{t/2} 3\tau_{12} \mathrm{d}3\,, \tag{2.89}$$

or all three relations combined in matrix notation:

$$\begin{bmatrix} M_1^{\mathrm{n}} \\ M_2^{\mathrm{n}} \\ M_{12}^{\mathrm{n}} \end{bmatrix} = \int_{-t/2}^{t/2} 3 \begin{bmatrix} \sigma_1 \\ \sigma_{12} \\ \sigma_{12} \end{bmatrix} \mathrm{d}3\,. \tag{2.90}$$

Furthermore, the shear forces per unit length are calculated in the following way:

$$Q_1^{\mathrm{n}} = \frac{Q_1}{\mathrm{d}2} = \int_{-t/2}^{t/2} \tau_{13} \mathrm{d}3\,, \tag{2.91}$$

$$Q_2^{\mathrm{n}} = \frac{Q_2}{\mathrm{d}1} = \int_{-t/2}^{t/2} \tau_{23} \mathrm{d}3\,, \tag{2.92}$$

or both relations combined in matrix notation:

$$\begin{bmatrix} Q_1^{\mathrm{n}} \\ Q_2^{\mathrm{n}} \end{bmatrix} = \int_{-t/2}^{t/2} \begin{bmatrix} \tau_{13} \\ \tau_{23} \end{bmatrix} \mathrm{d}3\,. \tag{2.93}$$

It should be noted that a slightly different notation when compared to the beam problems is used here. The bending moment around the 2-axis is now called M_1^{n} (which directly corresponds to the causing stress σ_1), while in the beam notation it was M_2; see Öchsner (2020a). Nevertheless, the orientation remains the same. The shear force, which was in the case of the beams given as Q_3, is now either Q_1^{n} or Q_2^{n}.

Thus, in the case of this plate notation, the index refers rather to the plane (check the surface normal vector) in which the corresponding resultant (vector) is located.

The equilibrium condition will be determined in the following for the vertical forces. Assuming that the distributed force is constant ($q_3(1, 2) \rightarrow q_3$) and that forces in the direction of the positive 3-axis are considered positive, it results as follows:

$$-Q_1^{\mathrm{n}}(1)\mathrm{d}2 - Q_2^{\mathrm{n}}(2)\mathrm{d}1 + Q_1^{\mathrm{n}}(1+\mathrm{d}1)\mathrm{d}2 + Q_2^{\mathrm{n}}(2+\mathrm{d}2)\mathrm{d}1 + q_3\mathrm{d}1\mathrm{d}2 = 0\,. \quad (2.94)$$

Expanding the shear forces at $1+\mathrm{d}1$ and $2+\mathrm{d}2$ in a Taylor's series of first order, meaning

$$Q_1^{\mathrm{n}}(1+\mathrm{d}1) \approx Q_1^{\mathrm{n}}(1) + \frac{\partial Q_1^{\mathrm{n}}}{\partial 1}\mathrm{d}1\,, \quad (2.95)$$

$$Q_2^{\mathrm{n}}(2+\mathrm{d}2) \approx Q_2^{\mathrm{n}}(2) + \frac{\partial Q_2^{\mathrm{n}}}{\partial 2}\mathrm{d}2\,, \quad (2.96)$$

Equation (2.94) results in

$$\frac{\partial Q_1^{\mathrm{n}}}{\partial 1}\mathrm{d}1\mathrm{d}2 + \frac{\partial Q_2^{\mathrm{n}}}{\partial 2}\mathrm{d}2\mathrm{d}1 + q_3\mathrm{d}1\mathrm{d}2 = 0\,, \quad (2.97)$$

or alternatively after simplification as

$$\frac{\partial Q_1^{\mathrm{n}}}{\partial 1} + \frac{\partial Q_2^{\mathrm{n}}}{\partial 2} + q_3 = 0\,. \quad (2.98)$$

The equilibrium of moments around the reference axis at $1+\mathrm{d}1$ (positive if the moment vector is pointing in positive 2-axis) gives

$$\begin{aligned} &M_1^{\mathrm{n}}(1+\mathrm{d}1)\mathrm{d}2 - M_1^{\mathrm{n}}(1)\mathrm{d}2 + M_{21}^{\mathrm{n}}(2+\mathrm{d}2)\mathrm{d}1 - M_{21}^{\mathrm{n}}\mathrm{d}1 \\ &- Q_2^{\mathrm{n}}(2)\mathrm{d}1\tfrac{\mathrm{d}1}{2} + Q_2^{\mathrm{n}}(2+\mathrm{d}2)\mathrm{d}1\tfrac{\mathrm{d}1}{2} - Q_1^{\mathrm{n}}(1)\mathrm{d}2\mathrm{d}1 + q_3\mathrm{d}1\mathrm{d}2\tfrac{\mathrm{d}1}{2} = 0\,. \end{aligned} \quad (2.99)$$

Expanding the stress resultants at $1+\mathrm{d}1$ and $2+\mathrm{d}2$ into a Taylor's series of first order, meaning

$$M_1^{\mathrm{n}}(1+\mathrm{d}1) = M_1^{\mathrm{n}}(1) + \frac{\partial M_1^{\mathrm{n}}}{\partial 1}\mathrm{d}1\,, \quad (2.100)$$

$$M_{21}^{\mathrm{n}}(2+\mathrm{d}2) = M_{21}^{\mathrm{n}}(2) + \frac{\partial M_{21}^{\mathrm{n}}}{\partial 2}\mathrm{d}2\,, \quad (2.101)$$

$$Q_2^{\mathrm{n}}(2+\mathrm{d}2) = Q_2^{\mathrm{n}}(2) + \frac{\partial Q_2^{\mathrm{n}}}{\partial 2}\mathrm{d}2\,, \quad (2.102)$$

Equation (2.99) results in

$$\frac{\partial M_1^{\mathrm{n}}}{\partial 1}\mathrm{d}1\mathrm{d}2+\frac{\partial M_{21}^{\mathrm{n}}}{\partial 2}\mathrm{d}2\mathrm{d}1+\frac{\partial Q_2^{\mathrm{n}}}{\partial 2}\mathrm{d}2\mathrm{d}1\frac{\mathrm{d}1}{2}-Q_1^{\mathrm{n}}(1)\mathrm{d}2\mathrm{d}1+q_3\mathrm{d}1\mathrm{d}2\frac{\mathrm{d}1}{2}=0\,. \tag{2.103}$$

Seeing that the terms of third order (d1d2d1) are considered as infinitesimally small and because of $M_{21}^{\mathrm{n}}=M_{12}^{\mathrm{n}}$, finally, it results as follows:

$$\frac{\partial M_1^{\mathrm{n}}}{\partial 1}+\frac{\partial M_{12}^{\mathrm{n}}}{\partial 2}-Q_1^{\mathrm{n}}=0\,. \tag{2.104}$$

In a similar way, the equilibrium of moments around the reference axis at $2+\mathrm{d}2$ finally gives

$$\frac{\partial M_2^{\mathrm{n}}}{\partial 2}+\frac{\partial M_{12}^{\mathrm{n}}}{\partial 1}-Q_2^{\mathrm{n}}=0\,. \tag{2.105}$$

Thus, the three equilibrium equations can be summarized as follows:

$$\frac{\partial Q_1^{\mathrm{n}}}{\partial 1}+\frac{\partial Q_2^{\mathrm{n}}}{\partial 2}+q_3=0\,, \tag{2.106}$$

$$\frac{\partial M_1^{\mathrm{n}}}{\partial 1}+\frac{\partial M_{12}^{\mathrm{n}}}{\partial 2}-Q_1^{\mathrm{n}}=0\,, \tag{2.107}$$

$$\frac{\partial M_2^{\mathrm{n}}}{\partial 2}+\frac{\partial M_{12}^{\mathrm{n}}}{\partial 1}-Q_2^{\mathrm{n}}=0\,. \tag{2.108}$$

Let us recall here that the following equilibrium equations for the Euler-Bernoulli beam can be obtained Öchsner (2021):

$$\frac{\mathrm{d}M_2(1)}{\mathrm{d}1}=Q_3(1)\,,\quad \frac{\mathrm{d}^2M_2(1)}{\mathrm{d}1^2}=\frac{\mathrm{d}Q_3(1)}{\mathrm{d}1}=-q_3\,. \tag{2.109}$$

Rearranging Eqs. (2.107) and (2.108) for Q^{n} and introducing them in Eq. (2.106) finally gives the combined equilibrium equation as

$$\frac{\partial^2 M_1^{\mathrm{n}}}{\partial 1^2}+2\frac{\partial^2 M_{12}}{\partial 1\partial 2}+\frac{\partial^2 M_2^{\mathrm{n}}}{\partial 2^2}+q_3=0\,. \tag{2.110}$$

The last equation can be written in matrix notation as

$$\begin{bmatrix}\frac{\partial^2}{\partial 1^2} & \frac{\partial^2}{\partial 2^2} & \frac{2\partial^2}{\partial 1\partial 2}\end{bmatrix}\begin{bmatrix}M_1^{\mathrm{n}}\\ M_2^{\mathrm{n}}\\ M_{12}^{\mathrm{n}}\end{bmatrix}+q_3=0\,, \tag{2.111}$$

or symbolically as

$$\boldsymbol{\mathcal{L}}_2^{\mathrm{T}}\boldsymbol{M}^{\mathrm{n}}+q_3=0\,. \tag{2.112}$$

Equations (2.107) and (2.108) can be rearranged to obtain a relationship between the moments and shear forces similar to Eq. $(2.109)_1$:

$$\mathcal{L}_1^{\mathrm{T}} \boldsymbol{M}^{\mathrm{n}} = \boldsymbol{Q}^{\mathrm{n}} \,, \tag{2.113}$$

where the first-order differential operator matrix $\mathcal{L}_1$ is given by Eqs. (2.79) and (2.80).

By combining Eqs. (2.85)–(2.86) and (2.111)–(2.112), the following equilibrium description for a *superposed element* is obtained

$$\left[\begin{array}{cccccc} \frac{\partial}{\partial 1} & 0 & \frac{\partial}{\partial 2} & 0 & 0 & 0 \\ 0 & \frac{\partial}{\partial 2} & \frac{\partial}{\partial 1} & 0 & 0 & 0 \\ \hline 0 & 0 & 0 & \frac{\partial^2}{\partial 1^2} & \frac{\partial^2}{\partial 2^2} & \frac{2\partial^2}{\partial 1 \partial 2} \end{array}\right] \left[\begin{array}{c} N_1^{\mathrm{n}} \\ N_2^{\mathrm{n}} \\ N_{12}^{\mathrm{n}} \\ \hline M_1^{\mathrm{n}} \\ M_2^{\mathrm{n}} \\ M_{12}^{\mathrm{n}} \end{array}\right] + \left[\begin{array}{c} q_1 \\ q_2 \\ \hline q_3 \end{array}\right] = \left[\begin{array}{c} 0 \\ 0 \\ \hline 0 \end{array}\right] , \tag{2.114}$$

or in symbolic notation:

$$\mathcal{L}^{\mathrm{T}} \boldsymbol{s}^{\mathrm{n}} + \boldsymbol{q} = \boldsymbol{0} \,, \tag{2.115}$$

where $\boldsymbol{s}^{\mathrm{n}}$ is the column matrix of the stress resultants per unit length (generalized stresses per unit length).

Let us note here that the derivation of the stress resultants N_i^{n} and M_i^{n} as given in Eqs. (2.84) and (2.90) must be revised for the combined case. This comes from the fact that the total strain in the form $\boldsymbol{\varepsilon} = \boldsymbol{\varepsilon}^0 + 3\boldsymbol{\kappa}$ (see Eq. (2.19)) must be used in the derivation:

$$\begin{bmatrix} N_1^{\mathrm{n}} \\ N_2^{\mathrm{n}} \\ N_{12}^{\mathrm{n}} \end{bmatrix} = \int_{-t/2}^{t/2} \boldsymbol{\sigma} \mathrm{d}3 \overset{(2.32)}{=} \int_{-t/2}^{t/2} \boldsymbol{C}\boldsymbol{\varepsilon} \mathrm{d}3 \overset{(2.19)}{=} \int_{-t/2}^{t/2} \boldsymbol{C} \left(\boldsymbol{\varepsilon}^0 + 3\boldsymbol{\kappa}\right) \mathrm{d}3 \tag{2.116}$$

$$= \int_{-t/2}^{t/2} \boldsymbol{C}\boldsymbol{\varepsilon}^0 \mathrm{d}3 + \int_{-t/2}^{t/2} \boldsymbol{C} 3\boldsymbol{\kappa} \mathrm{d}3 = \boldsymbol{C}\boldsymbol{\varepsilon}^0 [3]_{-t/2}^{t/2} + \boldsymbol{C}\boldsymbol{\kappa} [3^2/2]_{-t/2}^{t/2}$$

$$= t\boldsymbol{C}\boldsymbol{\varepsilon}^0 + \frac{t^2}{4} \boldsymbol{C}\boldsymbol{\kappa} \,. \tag{2.117}$$

A corresponding derivation for the internal bending moments gives

$$\begin{bmatrix} M_1^{\mathrm{n}} \\ M_2^{\mathrm{n}} \\ M_{12}^{\mathrm{n}} \end{bmatrix} = \int\limits_{-t/2}^{t/2} z\boldsymbol{\sigma}\mathrm{d}z \overset{(2.32)}{=} \int\limits_{-t/2}^{t/2} z\boldsymbol{C}\boldsymbol{\varepsilon}\mathrm{d}z \overset{(2.19)}{=} \int\limits_{-t/2}^{t/2} z\boldsymbol{C}\left(\boldsymbol{\varepsilon}^0 + z\boldsymbol{\kappa}\right)\mathrm{d}z \tag{2.118}$$

$$= \int\limits_{-t/2}^{t/2} \boldsymbol{C}\boldsymbol{\varepsilon}^0 z\mathrm{d}z + \int\limits_{-t/2}^{t/2} \boldsymbol{C}z^2\boldsymbol{\kappa}\mathrm{d}z = \boldsymbol{C}\boldsymbol{\varepsilon}^0[z^2/2]_{-t/2}^{t/2} + \boldsymbol{C}\boldsymbol{\kappa}[z^3/3]_{-t/2}^{t/2}$$

$$= \frac{t^2}{4}\boldsymbol{C}\boldsymbol{\varepsilon}^0 + \frac{t^2 3}{12}\boldsymbol{C}\boldsymbol{\kappa}\,. \tag{2.119}$$

Equations (2.117) and (2.119) can be combined to obtain a single matrix representation:

$$\begin{bmatrix} N_1^{\mathrm{n}} \\ N_2^{\mathrm{n}} \\ N_{12}^{\mathrm{n}} \\ \hline M_1^{\mathrm{n}} \\ M_2^{\mathrm{n}} \\ M_{12}^{\mathrm{n}} \end{bmatrix} = \left[\begin{array}{c|c} \underbrace{t\,\boldsymbol{C}}_{\boldsymbol{A}} & \underbrace{\frac{t^2}{4}\boldsymbol{C}}_{\boldsymbol{B}} \\ \hline \underbrace{\frac{t^2}{4}\boldsymbol{C}}_{\boldsymbol{B}} & \underbrace{\frac{t^3}{12}\boldsymbol{C}}_{\boldsymbol{D}} \end{array}\right] \begin{bmatrix} \varepsilon_1^0 \\ \varepsilon_2^0 \\ \gamma_{12}^0 \\ \hline \kappa_1 \\ \kappa_2 \\ \kappa_{12} \end{bmatrix}, \tag{2.120}$$

or in symbolic notation:

$$\boldsymbol{s}^{\mathrm{n}} = \boldsymbol{C}^*\boldsymbol{e}\,, \tag{2.121}$$

where $\boldsymbol{C}^*$ is the generalized elasticity matrix.

2.2.4 Partial Differential Equations

For a *plane elasticity element*, introducing the constitutive equation according to (2.32) in the equilibrium equation (2.80) gives

$$\boldsymbol{\mathcal{L}}_1^{\mathrm{T}}\boldsymbol{C}\boldsymbol{\varepsilon} + \boldsymbol{b} = \boldsymbol{0}\,. \tag{2.122}$$

Introducing the kinematics relations in the last equation according to (2.3) finally gives the Lamé-Navier equations:

$$\boldsymbol{\mathcal{L}}_1^{\mathrm{T}}\boldsymbol{C}\boldsymbol{\mathcal{L}}_1\boldsymbol{u}_{1,2} + \boldsymbol{b} = \boldsymbol{0}\,. \tag{2.123}$$

Alternatively, the displacements may be substituted and the differential equations are obtained in terms of stresses. This formulation is known as the Beltrami-Michell equations. If the body forces vanish ($\boldsymbol{b} = \boldsymbol{0}$), the partial differential equations in terms of stresses are called the Beltrami equations.

For a *classical plate element*, let us combine the three equations for the resulting moments according to Eqs. (2.87)–(2.89) in matrix notation as

$$\boldsymbol{M}^{\mathrm{n}} = \begin{bmatrix} M_1^{\mathrm{n}} \\ M_2^{\mathrm{n}} \\ M_{12}^{\mathrm{n}} \end{bmatrix} = \int_{-t/2}^{t/2} 3 \begin{bmatrix} \sigma_1 \\ \sigma_2 \\ \tau_{12} \end{bmatrix} \mathrm{d}3 = \int_{-t/2}^{t/2} 3\boldsymbol{\sigma}\mathrm{d}3\,. \tag{2.124}$$

Introducing Hooke's law (2.32) and the kinematics relation (2.15) gives for a constant elasticity matrix $\boldsymbol{C}$

$$\boldsymbol{M}^{\mathrm{n}} = -\int_{-t/2}^{t/2} 3^2 \boldsymbol{C}\boldsymbol{\mathcal{L}}_2 u_3 \mathrm{d}3 = -\boldsymbol{C}\boldsymbol{\mathcal{L}}_2 u_3 \underbrace{\int_{-t/2}^{t/2} 3^2 \mathrm{d}3}_{\frac{t^3}{12}} = -\frac{t^3}{12}\boldsymbol{C}\boldsymbol{\mathcal{L}}_2 u_3\,. \tag{2.125}$$

Using the kinematics relation in the curvature form (see Eq. (2.15)), it can be stated that

$$\boldsymbol{M}^{\mathrm{n}} = \frac{t^3}{12}\boldsymbol{C}\boldsymbol{\kappa}\,. \tag{2.126}$$

Introducing the moment-displacement relation (2.125) in the equilibrium equation (2.112) results in the plate bending differential equation in the form:

$$\boldsymbol{\mathcal{L}}_2^{\mathrm{T}}\left(\frac{t^3}{12}\boldsymbol{C}\boldsymbol{\mathcal{L}}_2 u_3\right) - q_3 = 0\,. \tag{2.127}$$

Assuming *isotropic* material behavior and the definitions for $\boldsymbol{\mathcal{L}}_2$ and $\boldsymbol{C}$ given in Eqs. (2.111) and (2.23), the following classical form of the plate bending differential equation can be obtained:

$$\frac{Et^3}{12(1-\nu^2)}\left(\frac{\partial^4 u_3}{\partial 1^4} + 2\frac{\partial^4 u_3}{\partial 1^2 \partial 2^2} + \frac{\partial^4 u_3}{\partial 2^4}\right) = q_3\,. \tag{2.128}$$

Let us recall here that the following partial differential equation for the Euler-Bernoulli beam can be obtained under the assumption of *isotropic* material behavior:

$$EI_2 \frac{\mathrm{d}^4 u_3(x)}{\mathrm{d}1^4} = q_3(x)\,, \tag{2.129}$$

where q_3 is now defined as a load per unit length.

Let us look in the following on the *combined plane elasticity and classical plate element*. Starting from the equilibrium equation as given in Eq. (2.114) and inserting Eq. (2.120), i.e., the equation which is based on the definition of the stress resultants under considering the constitutive equation and the relation for the sum of the strain components, gives with the kinematics relation Eq. (2.20) the following set of partial differential equations:

$$
\begin{bmatrix}
\frac{\partial}{\partial 1} & 0 & \frac{\partial}{\partial 2} & 0 & 0 & 0 \\
0 & \frac{\partial}{\partial 2} & \frac{\partial}{\partial 1} & 0 & 0 & 0 \\
0 & 0 & 0 & \frac{\partial^2}{\partial 1^2} & \frac{\partial^2}{\partial 2^2} & \frac{2\partial^2}{\partial 1 \partial 2}
\end{bmatrix}
\left[\begin{array}{c|c}
\underbrace{t\,\boldsymbol{C}}_{\boldsymbol{A}} & \underbrace{\frac{t^2}{4}\boldsymbol{C}}_{\boldsymbol{B}} \\ \hline
\underbrace{\frac{t^2}{4}\boldsymbol{C}}_{\boldsymbol{B}} & \underbrace{\frac{t^3}{12}\boldsymbol{C}}_{\boldsymbol{D}}
\end{array}\right]
\begin{bmatrix}
\frac{\partial}{\partial 1} & 0 & 0 \\
0 & \frac{\partial}{\partial 2} & 0 \\
\frac{\partial}{\partial 1} & \frac{\partial}{\partial 2} & 0 \\
0 & 0 & -\frac{\partial^2}{\partial 1^2} \\
0 & 0 & -\frac{\partial^2}{\partial 2^2} \\
0 & 0 & -\frac{2\partial^2}{\partial 1 \partial 2}
\end{bmatrix}
\begin{bmatrix} u_1 \\ u_2 \\ u_3 \end{bmatrix} +
$$

$$
+ \begin{bmatrix} q_1 \\ q_2 \\ q_3 \end{bmatrix} = \begin{bmatrix} 0 \\ 0 \\ 0 \end{bmatrix}, \tag{2.130}
$$

or

$$
\begin{bmatrix}
\frac{\partial}{\partial 1} & 0 & \frac{\partial}{\partial 2} & 0 & 0 & 0 \\
0 & \frac{\partial}{\partial 2} & \frac{\partial}{\partial 1} & 0 & 0 & 0 \\
0 & 0 & 0 & \frac{\partial^2}{\partial 1^2} & \frac{\partial^2}{\partial 2^2} & \frac{2\partial^2}{\partial 1 \partial 2}
\end{bmatrix}
\left[\begin{array}{ccc|ccc}
A_{11} & A_{12} & A_{14} & B_{11} & B_{12} & B_{14} \\
A_{12} & A_{22} & A_{24} & B_{12} & B_{22} & B_{24} \\
A_{14} & A_{24} & A_{24} & B_{44} & B_{24} & B_{44} \\ \hline
B_{11} & B_{12} & B_{14} & D_{11} & D_{12} & D_{14} \\
B_{12} & B_{22} & B_{24} & D_{12} & D_{22} & D_{24} \\
B_{14} & B_{24} & B_{44} & D_{14} & D_{24} & D_{44}
\end{array}\right]
\begin{bmatrix}
\frac{\partial}{\partial 1} & 0 & 0 \\
0 & \frac{\partial}{\partial 2} & 0 \\
\frac{\partial}{\partial 1} & \frac{\partial}{\partial 2} & 0 \\
0 & 0 & -\frac{\partial^2}{\partial 1^2} \\
0 & 0 & -\frac{\partial^2}{\partial 2^2} \\
0 & 0 & -\frac{2\partial^2}{\partial 1 \partial 2}
\end{bmatrix}
$$

$$
\begin{bmatrix} u_1 \\ u_2 \\ u_3 \end{bmatrix} + \begin{bmatrix} q_1 \\ q_2 \\ q_3 \end{bmatrix} = \begin{bmatrix} 0 \\ 0 \\ 0 \end{bmatrix}, \tag{2.131}
$$

or in symbolic notation:

$$
\boldsymbol{\mathcal{L}}^{\mathrm{T}} \boldsymbol{C}^{*} \boldsymbol{\mathcal{L}}' \boldsymbol{u} + \boldsymbol{q} = \boldsymbol{0}\,. \tag{2.132}
$$

Once the solution of the displacements $\boldsymbol{u}$ in Eq. (2.132) is obtained, for example, based on the finite element method Öchsner (2020a) or the finite difference method

Öchsner (2020b), the kinematics relationship (2.20) provides the strains from which the constitutive equation (2.31) allows the calculation of the stresses.

2.2.5 *Failure Criteria*

The following subsections present some of the common orthotropic failure criteria. The corresponding criteria for isotropic material behavior can be found, for example, in Öchsner (2016).

2.2.5.1 Maximum Stress Criterion

This criterion assumes that there is no failure, i.e., no fiber failure in the *1*-direction, no transversal stress failure in the *2*-direction, and no in-plane shear failure, as long as the stress components are in the following limits of the corresponding strength values k:

$$k_{1\mathrm{c}} < \sigma_1 < k_{1\mathrm{t}}\,, \tag{2.133}$$

$$k_{2\mathrm{c}} < \sigma_2 < k_{2\mathrm{t}}\,, \tag{2.134}$$

$$|\sigma_{12}| < k_{12\mathrm{s}}\,, \tag{2.135}$$

which assumes that there is no interaction between the different stress components. Some limitations of this criterion are discussed in Tsai (1968).

2.2.5.2 Maximum Strain Criterion

The maximum strain criterion is similar to the maximum stress criterion and assumes no interaction between the strain components. No failure occurs as long as the strain components are in the following limits:

$$k^{\varepsilon}_{1\mathrm{c}} < \varepsilon_1 < k^{\varepsilon}_{1\mathrm{t}}\,, \tag{2.136}$$

$$k^{\varepsilon}_{2\mathrm{c}} < \varepsilon_2 < k^{\varepsilon}_{2\mathrm{t}}\,, \tag{2.137}$$

$$|\gamma_{12}| < k^{\varepsilon}_{12\mathrm{s}}\,, \tag{2.138}$$

where the k^{ε} represent the corresponding failure strains. Some limitations of this criterion are discussed in Tsai (1968).

2.2.5.3 Tsai-Hill Criterion

The Tsai-Hill criterion is based on Hill's three-dimensional yield criterion for orthotropic materials which was applied by Tsai to laminae Hill (1950), Tsai (1968). This criterion can be stated for a plane stress state as follows:

$$\frac{\sigma_1^2}{k_1^2} - \frac{\sigma_1\sigma_2}{k_1^2} + \frac{\sigma_2^2}{k_2^2} + \frac{\sigma_{12}^2}{k_{12s}^2} < 1\,, \tag{2.139}$$

where $k_1 = k_{1t}$ for $\sigma_1 > 0$ or $k_1 = k_{1c}$ for $\sigma_1 < 0$. The same convention holds for direction 2. This criterion, which accounts for an interaction of the different stress components, showed a good agreement between the theoretical prediction and experimental values for E-glass-epoxy laminae at different rotational angles Tsai (1968).

2.2.5.4 Tsai-Wu Criterion

The Tsai-Wu criterion can be expressed for orthotropic materials under plane stress states as follows Tsai and Wu (1971):

$$a_1\sigma_1 + a_2\sigma_2 + a_{11}\sigma_1^2 + a_{22}\sigma_2^2 + a_{12}\sigma_{12}^2 + 2a_{1,2}\sigma_1\sigma_2 < 1\,, \tag{2.140}$$

where the coefficients a can be related to the classical failure stresses as follows Kaw (2006), Altenbach et al. (2018):

$$a_1 = \frac{1}{k_{1t}} + \frac{1}{k_{1c}}\,, \tag{2.141}$$

$$a_2 = \frac{1}{k_{2t}} + \frac{1}{k_{2c}}\,, \tag{2.142}$$

$$a_{11} = -\frac{1}{k_{1t}k_{1c}}\,, \tag{2.143}$$

$$a_{22} = -\frac{1}{k_{2t}k_{2c}}\,, \tag{2.144}$$

$$a_{12} = \frac{1}{k_{12s}^2}\,, \tag{2.145}$$

$$a_{1,2} \approx -\frac{1}{2}\sqrt{\frac{1}{k_{1t}k_{1c}} \times \frac{1}{k_{2t}k_{2c}}}\,. \tag{2.146}$$

Excellent agreement between the Tsai-Wu criterion and experimental data for boron-epoxy is reported in Byron Pipes and Cole (1973).

2.3 Macromechanics of a Laminate

Let us consider in the following a laminate, which is composed of n layers; see Fig. 2.11. Each layer k is a single lamina with its own orientation expressed in a local or lamina-specific coordinate system $(1_k, 2_k, 3_k)$. Thus, the global or laminate-specific coordinate system (x, y, z) is used to describe the orientation of the laminate.

The height of each layer k $(1 \leq k \leq n)$ can be expressed based on the thickness coordinate as

$$t_k = z_k - z_{k-1}\,, \tag{2.147}$$

and the total height of the laminate results in

$$t = \sum_{k=1}^{n} t_k\,. \tag{2.148}$$

2.3.1 Generalized Stress-Strain Relationship

Let us focus in the following on the evaluation of the stress resultants as introduced in Eqs. (2.116) and (2.118) for a single lamina. The internal normal forces can be expressed in the laminate-specific coordinate system (x, y, z) as

$$\begin{bmatrix} N_x^{\mathrm{n}} \\ N_y^{\mathrm{n}} \\ N_{xy}^{\mathrm{n}} \end{bmatrix} = \int_{-t/2}^{t/2} \boldsymbol{\sigma} \mathrm{d}z = \sum_{k=1}^{n} \int_{z_{k-1}}^{z_k} \boldsymbol{\sigma}_k \mathrm{d}\hat{z} = \sum_{k=1}^{n} \int_{z_{k-1}}^{z_k} \overline{\boldsymbol{C}}_k \boldsymbol{\varepsilon}_k \mathrm{d}\hat{z} \tag{2.149}$$

$$= \sum_{k=1}^{n} \int_{z_{k-1}}^{z_k} \overline{\boldsymbol{C}}_k \left(\boldsymbol{\varepsilon}^0 + \hat{z}\boldsymbol{\kappa}\right) \mathrm{d}\hat{z} \tag{2.150}$$

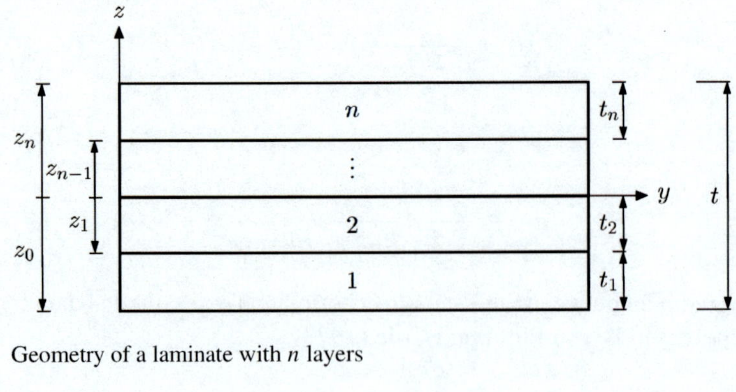

Fig. 2.11 Geometry of a laminate with n layers

$$= \sum_{k=1}^{n} \left(\int_{z_{k-1}}^{z_k} \overline{\boldsymbol{C}}_k \boldsymbol{\varepsilon}^0 \mathrm{d}\hat{z} + \int_{z_{k-1}}^{z_k} \overline{\boldsymbol{C}}_k \hat{z} \boldsymbol{\kappa} \mathrm{d}\hat{z} \right) \tag{2.151}$$

$$= \sum_{k=1}^{n} \left(\underbrace{\overline{\boldsymbol{C}}_k \left(z_k - z_{k-1}\right)}_{\boldsymbol{A}_k} \boldsymbol{\varepsilon}^0 + \underbrace{\overline{\boldsymbol{C}}_k \frac{1}{2} \left((z_k)^2 - (z_{k-1})^2\right)}_{\boldsymbol{B}_k} \boldsymbol{\kappa} \right), \tag{2.152}$$

and the corresponding derivation for the internal bending moments:

$$\begin{bmatrix} M_x^{\mathrm{n}} \\ M_y^{\mathrm{n}} \\ M_{xy}^{\mathrm{n}} \end{bmatrix} = \int_{-t/2}^{t/2} z \boldsymbol{\sigma} \mathrm{d}z = \sum_{k=1}^{n} \int_{z_{k-1}}^{z_k} \boldsymbol{\sigma}_k \hat{z} \mathrm{d}\hat{z} = \sum_{k=1}^{n} \int_{z_{k-1}}^{z_k} \overline{\boldsymbol{C}}_k \boldsymbol{\varepsilon}_k \hat{z} \mathrm{d}\hat{z} \tag{2.153}$$

$$= \sum_{k=1}^{n} \int_{z_{k-1}}^{z_k} \overline{\boldsymbol{C}}_k \left(\boldsymbol{\varepsilon}^0 \hat{z} + \hat{z}^2 \boldsymbol{\kappa}\right) \mathrm{d}\hat{z} \tag{2.154}$$

$$= \sum_{k=1}^{n} \left(\int_{z_{k-1}}^{z_k} \overline{\boldsymbol{C}}_k \boldsymbol{\varepsilon}^0 \hat{z} \mathrm{d}\hat{z} + \int_{z_{k-1}}^{z_k} \overline{\boldsymbol{C}}_k \hat{z}^2 \boldsymbol{\kappa} \mathrm{d}\hat{z} \right) \tag{2.155}$$

$$= \sum_{k=1}^{n} \left(\underbrace{\overline{\boldsymbol{C}}_k \frac{1}{2} \left((z_k)^2 - (z_{k-1})^2\right)}_{\boldsymbol{B}_k} \boldsymbol{\varepsilon}^0 + \underbrace{\overline{\boldsymbol{C}}_k \frac{1}{3} \left((z_k)^3 - (z_{k-1})^3\right)}_{\boldsymbol{D}_k} \boldsymbol{\kappa} \right). \tag{2.156}$$

Equations (2.152) and (2.156) can be combined in a single matrix form to give

tension-shear coupling; tension-twist, bending-shear coupling

$$\begin{bmatrix} N_x^{\mathrm{n}} \\ N_y^{\mathrm{n}} \\ N_{xy}^{\mathrm{n}} \\ \hline M_x^{\mathrm{n}} \\ M_y^{\mathrm{n}} \\ M_{xy}^{\mathrm{n}} \end{bmatrix} = \left[\begin{array}{ccc|ccc} A_{11} & A_{12} & A_{14} & B_{11} & B_{12} & B_{14} \\ A_{12} & A_{22} & A_{24} & B_{12} & B_{22} & B_{24} \\ A_{14} & A_{24} & A_{24} & B_{44} & B_{24} & B_{44} \\ \hline B_{11} & B_{12} & B_{14} & D_{11} & D_{12} & D_{14} \\ B_{12} & B_{22} & B_{24} & D_{12} & D_{22} & D_{24} \\ B_{14} & B_{24} & B_{44} & D_{14} & D_{24} & D_{44} \end{array} \right] \begin{bmatrix} \varepsilon_x^0 \\ \varepsilon_y^0 \\ \gamma_{xy}^0 \\ \hline \kappa_x \\ \kappa_y \\ \kappa_{xy} \end{bmatrix}, \tag{2.157}$$

bending-tension coupling; bending-twist coupling

or more symbolically

$$\underbrace{\begin{bmatrix} \boldsymbol{N}^{\mathrm{n}} \\ \hline \boldsymbol{M}^{\mathrm{n}} \end{bmatrix}}_{\boldsymbol{s}} \underbrace{\left[\begin{array}{c|c} \boldsymbol{A} & \boldsymbol{B} \\ \hline \boldsymbol{B} & \boldsymbol{D} \end{array} \right]}_{\boldsymbol{C}^*} = \underbrace{\begin{bmatrix} \boldsymbol{\varepsilon}^0 \\ \hline \boldsymbol{\kappa} \end{bmatrix}}_{\boldsymbol{e}}, \tag{2.158}$$

where $\boldsymbol{s}$ is the column matrix of stress resultants (generalized stresses), $\boldsymbol{C}^*$ is the generalized elasticity matrix, and $\boldsymbol{e}$ is the column matrix of generalized strains. The corresponding submatrices are given as follows:

$$\boldsymbol{A} = \sum_{k=1}^{n} \boldsymbol{A}_k = \sum_{k=1}^{n} \overline{\boldsymbol{C}}_k \, (z_k - z_{k-1}) \,, \tag{2.159}$$

$$\boldsymbol{B} = \sum_{k=1}^{n} \boldsymbol{B}_k = \frac{1}{2}\sum_{k=1}^{n} \overline{\boldsymbol{C}}_k \left((z_k)^2 - (z_{k-1})^2\right) \,, \tag{2.160}$$

$$\boldsymbol{D} = \sum_{k=1}^{n} \boldsymbol{D}_k = \frac{1}{3}\sum_{k=1}^{n} \overline{\boldsymbol{C}}_k \left((z_k)^3 - (z_{k-1})^3\right) \,. \tag{2.161}$$

It should be noted here that $\boldsymbol{A}$ is called the extensional submatrix, $\boldsymbol{D}$ the bending submatrix, and $\boldsymbol{B}$ the coupling submatrix.

Equation (2.158) can be inverted to obtain the strains and curvatures as a function of the generalized stresses as

$$\underbrace{\left[\begin{array}{c}\boldsymbol{\varepsilon}^0 \\ \hline \boldsymbol{\kappa}\end{array}\right]}_{\boldsymbol{e}} = \left(\left[\begin{array}{c|c}\boldsymbol{A} & \boldsymbol{B} \\ \hline \boldsymbol{B} & \boldsymbol{D}\end{array}\right]\right)^{-1} \left[\begin{array}{c}\boldsymbol{N}^{\mathrm{n}} \\ \hline \boldsymbol{M}^{\mathrm{n}}\end{array}\right] = \underbrace{\left[\begin{array}{c|c}\boldsymbol{A}' & \boldsymbol{B}' \\ \hline (\boldsymbol{B}')^{\mathrm{T}} & \boldsymbol{D}'\end{array}\right]}_{(\boldsymbol{C}^*)^{-1}} \underbrace{\left[\begin{array}{c}\boldsymbol{N}^{\mathrm{n}} \\ \hline \boldsymbol{M}^{\mathrm{n}}\end{array}\right]}_{\boldsymbol{s}} , \tag{2.162}$$

where the submatrices are given as follows Chawla (1987), Altenbach et al. (2018):

$$\boldsymbol{A}' = \boldsymbol{A}^{-1} + \left(-\boldsymbol{A}^{-1}\boldsymbol{B}\right)\left(\boldsymbol{D} - \boldsymbol{B}\boldsymbol{A}^{-1}\boldsymbol{B}\right)^{-1}\left(-\boldsymbol{A}^{-1}\boldsymbol{B}\right)^{\mathrm{T}} \,, \tag{2.163}$$

$$\boldsymbol{B}' = \left(-\boldsymbol{A}^{-1}\boldsymbol{B}\right)\left(\boldsymbol{D} - \boldsymbol{B}\boldsymbol{A}^{-1}\boldsymbol{B}\right)^{-1} \,, \tag{2.164}$$

$$\boldsymbol{D}' = \left(\boldsymbol{D} - \boldsymbol{B}\boldsymbol{A}^{-1}\boldsymbol{B}\right)^{-1} \,. \tag{2.165}$$

As can be concluded from Eq. (2.157), there is a coupling between tension and shear ($A_{14} : N^{\mathrm{n}}_x \rightarrow \gamma^0_{xy}$; $A_{24} : N^{\mathrm{n}}_y \rightarrow \gamma^0_{xy}$), bending and tension ($B_{11}, B_{12} : M^{\mathrm{n}}_x \rightarrow \varepsilon^0_x, \varepsilon^0_y$; $B_{12}, B_{22} : M^{\mathrm{n}}_y \rightarrow \varepsilon^0_x, \varepsilon^0_y$), tension and twist ($B_{14} : N^{\mathrm{n}}_x \rightarrow \kappa_{xy}$; $B_{24} : N^{\mathrm{n}}_y \rightarrow \kappa_{xy}$), bending and shear ($B_{14} : M^{\mathrm{n}}_x \rightarrow \gamma^0_{xy}$; $B_{24} : M^{\mathrm{n}}_y \rightarrow \gamma^0_{xy}$), and bending and twist ($D_{14} : M^{\mathrm{n}}_x \rightarrow \kappa_{xy}$; $D_{24} : M^{\mathrm{n}}_y \rightarrow \kappa_{xy}$).

Based on the generalized strains obtained from Eq. (2.162), it is now possible to calculate the stresses in each layer k expressed in the x–y coordinate system:

$$\boldsymbol{\sigma}^k_{x,y}(z) = \overline{\boldsymbol{C}}_k \left(\boldsymbol{\varepsilon}^0 + z\boldsymbol{\kappa}\right) , \tag{2.166}$$

where the vertical coordinate z ranges for the kth layer in the following boundaries: $z_{k-1} \leq z \leq z_k$. The stresses may be evaluated at the bottom ($z = z_{k-1}$), middle

($z = (z_k + z_{k-1})/2$), or top ($z = z_k$) of each layer. Based on relation (2.36), we can transform the stress values into the *1–2* coordinate system (see Fig. 2.6):

$$\boldsymbol{\sigma}^k_{1,2} = \boldsymbol{T}^k_\sigma \boldsymbol{\sigma}^k_{x,y} \,. \tag{2.167}$$

The obtained stress values may serve for a subsequent failure analysis of layer k (see Sect. 2.3.3).

2.3.2 *Special Cases of Laminates*

The following section covers special cases of the generalized elasticity matrix $\boldsymbol{C}^*$ as provided in Eq. (2.157); see MIL-HDBK-17/2F (2002), Altenbach et al. (2018) for details.

- Symmetric laminates:

Symmetric laminates have a symmetric buildup sequence in regard to laminae (plies) orientations, thicknesses, and material properties about the middle surface; see Fig. 2.12 for some examples.

The generalized stiffness matrix (2.157) simplifies for symmetric laminates to the following form:

$$\boldsymbol{C}^* = \left[\begin{array}{ccc|ccc} A_{11} & A_{12} & A_{14} & 0 & 0 & 0 \\ A_{12} & A_{22} & A_{24} & 0 & 0 & 0 \\ A_{14} & A_{24} & A_{24} & 0 & 0 & 0 \\ \hline 0 & 0 & 0 & D_{11} & D_{12} & D_{14} \\ 0 & 0 & 0 & D_{12} & D_{22} & D_{24} \\ 0 & 0 & 0 & D_{14} & D_{24} & D_{44} \end{array}\right], \tag{2.168}$$

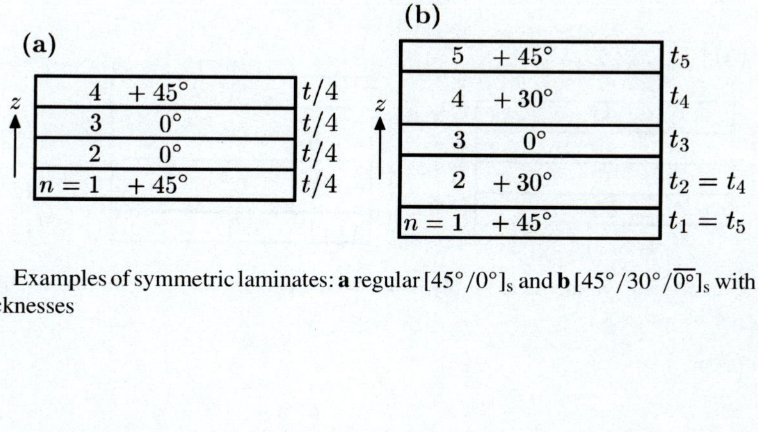

Fig. 2.12 Examples of symmetric laminates: **a** regular $[45°/0°]_s$ and **b** $[45°/30°/\overline{0°}]_s$ with different layer thicknesses

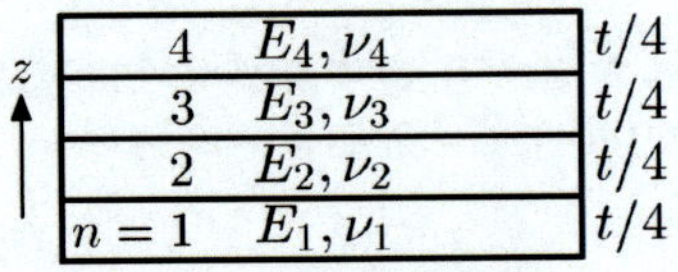

Fig. 2.13 Example of a symmetric laminate with isotropic layers

where we have now $\boldsymbol{B} = \boldsymbol{0}$.

• Symmetric laminates with isotropic layers:

Symmetric laminates with isotropic layers contain only isotropic laminae (plies), i.e., laminae where the elastic properties are independent of the direction and characterized by two independent properties (e.g., elasticity modulus and Poisson's ratio); see Fig. 2.13.

The generalized stiffness matrix (2.157) simplifies for symmetric laminates with isotropic layers to the following form:

$$\boldsymbol{C}^* = \left[\begin{array}{ccc|ccc} A_{11} & A_{12} & 0 & 0 & 0 & 0 \\ A_{12} & A_{11} & 0 & 0 & 0 & 0 \\ 0 & 0 & A_{24} & 0 & 0 & 0 \\ \hline 0 & 0 & 0 & D_{11} & D_{12} & 0 \\ 0 & 0 & 0 & D_{12} & D_{11} & 0 \\ 0 & 0 & 0 & 0 & 0 & D_{44} \end{array}\right] . \tag{2.169}$$

• Symmetric cross-ply laminates:

Symmetric cross-ply laminates only contain orthotropic laminae (plies) at right angles to one another; see Fig. 2.14 for some examples. Laminae may have pairwise (i.e., to ensure symmetry) different thicknesses and material properties.

The generalized stiffness matrix (2.157) simplifies for symmetric cross-ply laminates to the following form:

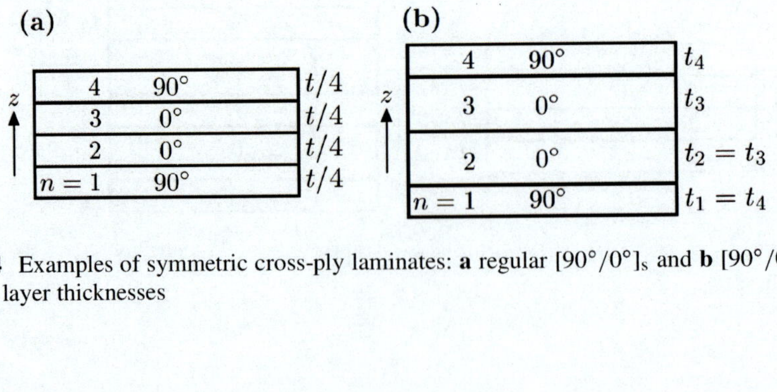

Fig. 2.14 Examples of symmetric cross-ply laminates: **a** regular $[90°/0°]_s$ and **b** $[90°/0°]_s$ with different layer thicknesses

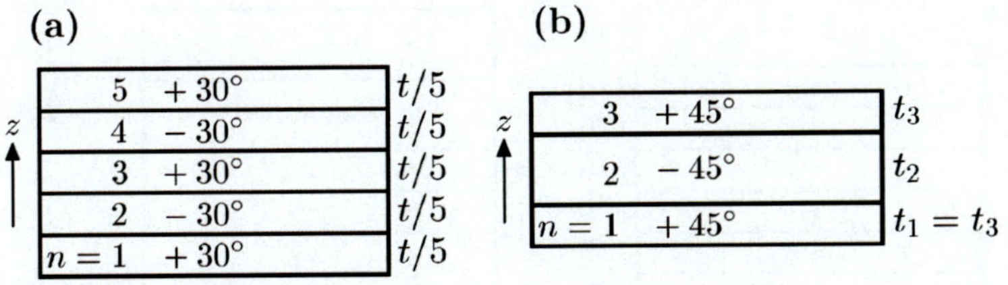

Fig. 2.15 Examples of symmetric angle-ply laminates: **a** regular $[30°/-30°/\overline{30°}]_s$ and **b** $[45°/\overline{-45°}]_s$ with different layer thicknesses

$$
\boldsymbol{C}^* = \left[\begin{array}{ccc|ccc}
A_{11} & A_{12} & 0 & 0 & 0 & 0 \\
A_{12} & A_{22} & 0 & 0 & 0 & 0 \\
0 & 0 & A_{24} & 0 & 0 & 0 \\
\hline
0 & 0 & 0 & D_{11} & D_{12} & 0 \\
0 & 0 & 0 & D_{12} & D_{22} & 0 \\
0 & 0 & 0 & 0 & 0 & D_{44}
\end{array}\right]. \tag{2.170}
$$

• Symmetric angle-ply laminates:

Symmetric angle-ply laminates only contain orthotropic laminae (plies) at which adjacent laminae have opposite orientation signs; see Fig. 2.15 for some examples. Laminae may have pair-wise (i.e., to ensure symmetry) different thicknesses and material properties.

The generalized stiffness matrix (2.157) simplifies for symmetric angle-ply laminates to the following form:

$$
\boldsymbol{C}^* = \left[\begin{array}{ccc|ccc}
A_{11} & A_{12} & A_{14} & 0 & 0 & 0 \\
A_{12} & A_{22} & A_{24} & 0 & 0 & 0 \\
A_{14} & A_{24} & A_{24} & 0 & 0 & 0 \\
\hline
0 & 0 & 0 & D_{11} & D_{12} & D_{14} \\
0 & 0 & 0 & D_{12} & D_{22} & D_{24} \\
0 & 0 & 0 & D_{14} & D_{24} & D_{44}
\end{array}\right]. \tag{2.171}
$$

It should be noted here that we obtain for the special case $[45°/\overline{-45°}]_s$ (see Fig. 2.15b) the further simplification $A_{22} = A_{11}$. However, $[30°/\overline{-30°}]_s$ yields $A_{22} \neq A_{11}$.

• Symmetric balanced laminates:

Symmetric angle-ply laminates only contain orthotropic laminae (plies) at which all identical laminae at angles other than 0° and 90° occur only in ± pairs (not necessarily

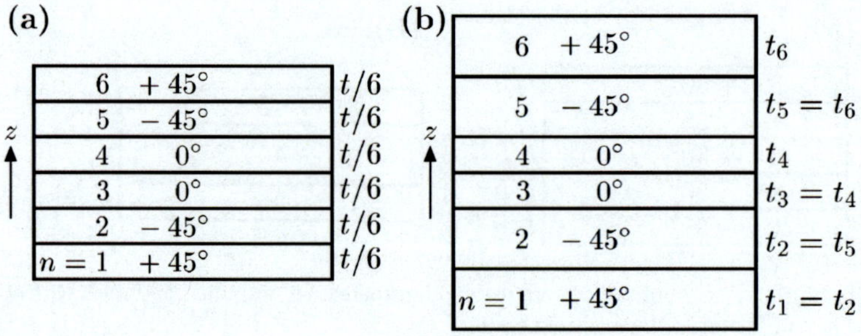

Fig. 2.16 Examples of symmetric balanced laminates: **a** regular $[45°/-45°/0°]_s$ and **b** $[45°/-45°/0°]_s$ with different layer thicknesses

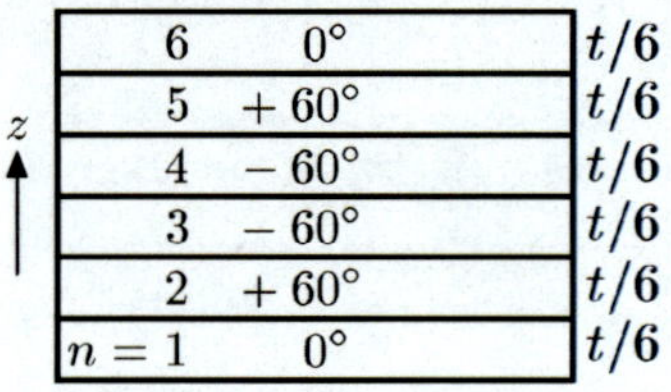

Fig. 2.17 Example of a quasi-isotropic symmetric laminate: $[0°/60°/-60°]_s$

adjacent); see Fig. 2.16 for some examples. These laminates only contain an even number of laminae.[6]

The generalized stiffness matrix (2.157) simplifies for symmetric balanced laminates to the following form:

$$\boldsymbol{C}^* = \left[\begin{array}{ccc|ccc} A_{11} & A_{12} & 0 & 0 & 0 & 0 \\ A_{12} & A_{22} & 0 & 0 & 0 & 0 \\ 0 & 0 & A_{24} & 0 & 0 & 0 \\ \hline 0 & 0 & 0 & D_{11} & D_{12} & D_{14} \\ 0 & 0 & 0 & D_{12} & D_{22} & D_{24} \\ 0 & 0 & 0 & D_{14} & D_{24} & D_{44} \end{array}\right]. \tag{2.172}$$

- Symmetric quasi-isotropic laminates:

Symmetric quasi-isotropic laminates are balanced and only contain orthotropic laminae (plies) for which a constitutive property of interest, at a given point, displays isotropic behavior in the *plane* of the laminates; see Fig. 2.17.

[6] Thus, symmetric angle-ply laminates do not belong to the group of symmetric balanced laminates.

The generalized stiffness matrix (2.157) simplifies for symmetric quasi-isotropic laminates to the following form:

$$C^* = \left[\begin{array}{ccc|ccc} A_{11} & A_{12} & 0 & 0 & 0 & 0 \\ A_{12} & A_{11} & 0 & 0 & 0 & 0 \\ 0 & 0 & A_{24} & 0 & 0 & 0 \\ \hline 0 & 0 & 0 & D_{11} & D_{12} & D_{14} \\ 0 & 0 & 0 & D_{12} & D_{22} & D_{24} \\ 0 & 0 & 0 & D_{14} & D_{24} & D_{44} \end{array}\right]. \tag{2.173}$$

It should be noted here that symmetric quasi-isotropic laminates require that all layers have the same thickness and the same orthotropic material properties. As an example, $t_1 = t_6$ but $t_2 = t_3 = t_4 = t_5 \neq t_1$ would yield $A_{11} \neq A_{22}$. In a similar way, $\text{Mat}_1 = \text{Mat}_6$ but $\text{Mat}_2 = \text{Mat}_3 = \text{Mat}_4 = \text{Mat}_5 \neq \text{Mat}_1$ would yield $A_{11} \neq A_{22}$. Thus, such laminates would be no longer quasi-isotropic.

2.3.3 Failure Analysis of Laminates

Let us summarize here the recommended steps for a calculation according to the classical laminate theory (internal normal forces N_i^{n} and internal bending moments M_i^{n} given); see Table 2.2 for the case that the internal normal forces and internal bending moments are given and Table 2.3 for the case that the generalized strains are given.

Once the stresses and/or strains in each lamina are known, the failure analysis can be performed as outlined in Sect. 2.2.5. To easily compare the different failure criteria, one may determine the so-called limit loads, i.e., to multiply the loads with the strength ratio R. In case of $R > 1$ there is no failure, whereas $R \leq 1$ results in the failure of the respective lamina. Thus, the conditions of the maximum stress criterion according to Eqs. (2.133)–(2.135) read

$$(k_{1\text{c}} <)\ R \times \sigma_1 < k_{1\text{t}}\,, \tag{2.174}$$

$$(k_{2\text{c}} <)\ R \times \sigma_2 < k_{2\text{t}}\,, \tag{2.175}$$

$$R \times |\sigma_{12}| < k_{12\text{s}}\,, \tag{2.176}$$

from which the load ratio R can be identified. The conditions of the maximum strain criterion according to Eqs. (2.136)–(2.138) read

$$k^{\varepsilon}_{1\text{c}} < R \times \varepsilon_1 < k^{\varepsilon}_{1\text{t}}\,, \tag{2.177}$$

$$(k^{\varepsilon}_{2\text{c}} <)\ R \times \varepsilon_2 < k^{\varepsilon}_{2\text{t}}\,, \tag{2.178}$$

$$R \times |\gamma_{12}| < k^{\varepsilon}_{12\text{s}}\,, \tag{2.179}$$

Table 2.2 Recommended steps for a calculation according to the classical laminate theory. Case: internal normal forces N_i^{n} and internal bending moments M_i^{n} given

Nr.	Steps to perform
①	Define for each lamina k the material $(E_{1,k};\ E_{2,k};\ \nu_{12,k};\ G_{12,k})$ and the geometrical $(z_k;\ z_{k-1};\ \alpha_k)$ properties
②	Define the column matrix of generalized stresses: $\boldsymbol{s} = \left[N_x^{\mathrm{n}}\ N_y^{\mathrm{n}}\ N_{xy}^{\mathrm{n}}\ \ M_x^{\mathrm{n}}\ M_y^{\mathrm{n}}\ M_{xy}^{\mathrm{n}}\right]^{\mathrm{T}}$
③	Calculate for each layer k the elasticity matrix $\boldsymbol{C}_k$ in the *1–2* lamina system according to Eq. (2.31). Transform each matrix according to Eq. (2.49) to obtain the elasticity matrix $\overline{\boldsymbol{C}}_k$ in the x–y laminate system
④	Calculate the submatrices $\boldsymbol{A}$, $\boldsymbol{B}$, and $\boldsymbol{C}$ according to Eqs. (2.159)–(2.161). Assemble the generalized elasticity matrix $\boldsymbol{C}^*$ according to Eq. (2.158)
⑤	Calculate the generalized compliance matrix $(\boldsymbol{C}^*)^{-1}$ based on Eqs. (2.163)–(2.165)
⑥	Calculate the generalized strains $\boldsymbol{e} = \left[\boldsymbol{\varepsilon}^0\ \boldsymbol{\kappa}\right]^{\mathrm{T}}$ according to Eq. (2.162)
⑦	Calculate the stresses in each layer k according to Eq. (2.166), $\boldsymbol{\sigma}^k_{x,y}(z) = \overline{\boldsymbol{C}}_k\left(\boldsymbol{\varepsilon}^0 + z\boldsymbol{\kappa}\right)$, in the x–y laminate system $(z_{k-1} \leq z \leq z_k)$. The strains in the x–y laminate system are obtained from $\left(\boldsymbol{\varepsilon}^0 + z\boldsymbol{\kappa}\right)$
⑧	Transform the stresses in each layer to the *1–2* lamina system according to Eq. (2.167): $\boldsymbol{\sigma}^k_{1,2} = \boldsymbol{T}^k_{\sigma}\boldsymbol{\sigma}^k_{x,y}$. The strains in the *1–2* lamina system are obtained from Eq. (2.33): $\boldsymbol{\varepsilon}^k_{1,2} = (\boldsymbol{C}_k)^{-1}\boldsymbol{\sigma}^k_{1,2}$

Table 2.3 Recommended steps for a calculation according to the classical laminate theory. Case: generalized strains $\boldsymbol{e} = \left[\varepsilon_x^0\ \varepsilon_y^0\ \gamma_{xy}^0\ \kappa_x\ \kappa_y\ \kappa_{xy}\right]^{\mathrm{T}}$ given

Nr.	Steps to perform
①	Define for each lamina k the material $(E_{1,k};\ E_{2,k};\ \nu_{12,k};\ G_{12,k})$ and the geometrical $(z_k;\ z_{k-1};\ \alpha_k)$ properties
②	Define the column matrix of generalized strains: $\boldsymbol{e} = \left[\varepsilon_x^0\ \varepsilon_y^0\ \gamma_{xy}^0\ \kappa_x\ \kappa_y\ \kappa_{xy}\right]^{\mathrm{T}}$
③	Calculate for each layer k the elasticity matrix $\boldsymbol{C}_k$ in the *1–2* lamina system according to Eq. (2.31). Transform each matrix according to Eq. (2.49) to obtain the elasticity matrix $\overline{\boldsymbol{C}}_k$ in the x–y laminate system
④	Calculate the submatrices $\boldsymbol{A}$, $\boldsymbol{B}$, and $\boldsymbol{C}$ according to Eqs. (2.159)–(2.161). Assemble the generalized elasticity matrix $\boldsymbol{C}^*$ according to Eq. (2.158)
⑤	Calculate the generalized stresses $\boldsymbol{s} = \left[\boldsymbol{N}^{\mathrm{n}}\ \boldsymbol{M}^{\mathrm{n}}\right]$ according to Eq. (2.157)
(⑥)	In case that the stresses and strains in each layer are required, go to steps ⑦ and ⑧ in Table 2.2

from which the load ratio R can be identified. The Tsai-Hill criterion according to Eq. (2.139) reads

$$\frac{(R \times \sigma_1)^2}{k_1^2} - \frac{(R \times \sigma_1)(R \times \sigma_2)}{k_1^2} + \frac{(R \times \sigma_2)^2}{k_2^2} + \frac{(R \times \sigma_{12})^2}{k_{12s}^2} < 1\,, \qquad (2.180)$$

from which the load ratio R can be identified. And finally, the Tsai-Wu criterion according to Eq. (2.140) reads

$$a_1(R \times \sigma_1) + a_2(R \times \sigma_2) + a_{11}(R \times \sigma_1)^2 + a_{22}(R \times \sigma_2)^2 + a_{12}(R \times \sigma_{12})^2 + 2a_{1,2}(R \times \sigma_1)(R \times \sigma_2) < 1\,. \qquad (2.181)$$

This is a classical quadratic equation in R, from which the value of R can be identified.

References

Altenbach H, Altenbach J, Kissing W (2018) Mechanics of composite structural elements. Springer Nature, Singapore

Altenbach H, Öchsner A (eds) (2020) Encyclopedia of continuum mechanics. Springer, Berlin

Ashton JE, Whitney JM (1970) Theory of laminated plates. Technomic Publishing, Stamford

Barbero EJ (1996) Finite element analysis of composite materials. CRC Press, Boca Raton

Blaauwendraad J (2010) Plates and FEM: surprises and pitfalls. Springer, Dordrecht

Chawla KK (1987) Composite materials: science and engineering. Springer, New York

Chen WF, Han DJ (1988) Plasticity for structural engineers. Springer, New York

Clyne TW, Hull D (2019) An introduction to composite materials. Cambridge University Press, Cambridge

Daniel IM, Ishai O (1994) Engineering mechanics of composite materials. Oxford University Press, New York

Dong SB, Pister KS, Taylor RL (1962) On the theory of laminated anisotropic shells and plates. J Aerosp Sci 29:696–975

Eschenauer H, Olhoff N, Schnell W (1997) Applied structural mechanics: fundamentals of elasticity, load-bearing structures, structural optimization. Springer, Berlin

Hill R (1950) The mathematical theory of plasticity. Oxford University Press, London

Kaw AK (2006) Mechanics of composite materials. CRC Press, Boca Raton

Jones RM (1975) Mechanics of composite materials. Scripta Book Co, Washington

Matthews FL, Davies GAO, Hitchings D, Soutis C (2000) Finite element modelling of composite materials and structures. Woodhead Publishing Limited, Cambridge

MIL-HDBK-17, 2F, (2002) Composite materials handbook-, vol 2. Polymer matrix composites materials properties, Department of Defence, USA

Ochoa OO, Reddy JN (1992) Finite element analysis of composite laminates. Kluwer Academic Publishers, Dordrecht

Öchsner A (2016) Continuum damage and fracture mechanics. Springer, Singapore

Öchsner A, Öchsner M (2018) A first introduction to the finite element analysis program MSC Marc/Mentat. Springer, Cham

Öchsner A (2020) Computational statics and dynamics: an introduction based on the finite element method. Springer, Singapore

Öchsner A (2020) Structural mechanics with a pen: a guide to solve finite difference problems. Springer, Cham
Öchsner A (2021) Classical beam theories of structural mechanics. Springer, Cham
Pipes RB, Cole BW (1973) On the off-axis strength test for anisotropic materials. J Compos Mater 7:246–256
Reddy JN (2006) An introduction to the finite element method. McGraw Hill, Singapore
Stravsky Y (1961) Bending and stretching of laminated aeolotropic plates. J Eng Mech Div-ASCE 87:31–56
Tenek LT, Argyris J (1998) Finite element analysis for composite structures. Springer, Dordrecht
Timoshenko S, Woinowsky-Krieger S (1959) Theory of plates and shells. McGraw-Hill Book Company, New York
Tsai SW (1968) Strength theories of filamentary structures. In: Schwartz RT, Schwartz HS (eds) Fundamental aspects of fiber reinforced plastic composites. Wiley Interscience, New York
Tsai SW, Wu EM (1971) A general theory of strength for anisotropic materials. J Compos Mater 5:58–80
Ventsel E, Krauthammer T (2001) Thin plates and shells: theory, analysis, and applications. Marcel Dekker, New York
Wang CM, Reddy JN, Lee KH (2000) Shear deformable beams and plates: relationships with classical solution. Elsevier, Oxford

Chapter 3
Composite Laminate Analysis Tool—CLAT

Abstract This chapter describes the programming approach for the Python-based computational tool, the so-called Composite Laminate Analysis Tool (CLAT). The choices for the different tools and libraries, particularly for our user interface, are explained in detail. The second part of this chapter covers the graphical appearance, functionality, and practical handling of CLAT.

3.1 Programming Approach

When creating web-based applications, the first question to be answered is on which framework the application will be developed. The answer depends on several criteria, including the technical requirements, planned applications of the final product, and the preferred development programming language. Our framework of choice for CLAT is Plotly Dash (https://dash.plotly.com/) which fully supports Python development and gives us the flexibility we need to provide a seamless experience to the user. We have leveraged the capabilities of Dash using third-party libraries such as Dash Bootstrap Components (https://dash-bootstrap-components.opensource.faculty.ai) and Dash Mantine Components (https://www.dash-mantine-components.com) which provide additional tools to the basic framework. We also use Plotly Open-Source Graphing Library for Python (https://plotly.com/python/) to plot the models and the results.

Here, we concisely cover the basic ideas and concepts deployed in creating CLAT to acquaint the reader with the program flow. For further development and extension of the current source code, the interested reader is highly encouraged to consult the tutorial materials provided in Plotly Dash's web portal. Like any other application with a graphical user interface, CLAT consists of two main parts: frontend and backend. The backend is responsible for the calculations done for the composite material and its layers, while the frontend provides a user interface to communicate the input data required for those calculations, and to demonstrate the outcome of the analyses.

A. Öchsner and R. Makvandi, *A Numerical Approach to the Classical Laminate Theory of Composite Materials*, Advanced Structured Materials 189,
https://doi.org/10.1007/978-3-031-32975-3_3

An object-oriented approach has been used in developing the backend of CLAT; there are two main classes, i.e., *Laminate* and *Lamina* . An object of *Laminate* class is a collection of *Lamina* objects. The relation between the two classes is rather a composition than inheritance; in other words, there is no parent-child relation between the two types of objects. While most of the class methods are exposed to the user to be called inside a developing environment, there are also private methods included, which are only supposed to be called inside class-level definitions. These methods are distinguished from their public counterparts using a "__" prefix (for instance, *__getBendingSubmatrix* method in the *Laminate* class definition).

The main Python file starts with the definitions of some functions used for the graphical user interface which is done by importing the *CLATHelpModule.py*, followed by initializing the data for a sample composite material and the corresponding loading conditions for the analysis. We use *Dash DataTables* (https://dash.plotly.com/datatable) to provide a tool to demonstrate and modify the details of the model. An introduction to the Python programming language can be found in the common literature (Pilgrim 2004; Lee 2011; Bernard 2016; Kalb 2016).

In Dash, *@app.callback* is used to handle events (an event happens, for instance, when a value inside a data table is modified). Each callback consists of input and output data which can be recognized by their tags. In a basic configuration, the input data can be provided in two ways; using *input* or *state* tags. The most crucial difference between the two is that an *input* triggers an event. In contrast, *state* data is only transmitted without being able to initiate an event. Simply put, a change in a *state* input won't cause any events to get triggered. However, having them as inputs guarantees that the following functions use an actual version of the object. The function name coming immediately after the app.callback decorators gives a hint on what that specific callback/event is supposed to do.

We have chosen the function and variable names to be self-explanatory to avoid writing unnecessary comments in the code. In general, it should be easy for the reader to go through the script line by line and understand the flow of the program; see Chap. 5.

3.2 Graphical Appearance, Functionality, and Operation

To run CLAT, the following program files should be located in the working directory of the computer:

- `main.py`: The main program code to drive CLAT.
- `Laminates_classes.py`: It contains the two main classes, i.e., *Laminate* and *Lamina* .
- `globalVars.py`: It contains the global variables.
- `CLATHelpModule`: It contains all the additional functions.
- `requirements.txt`: To install the current versions of the required libraries.

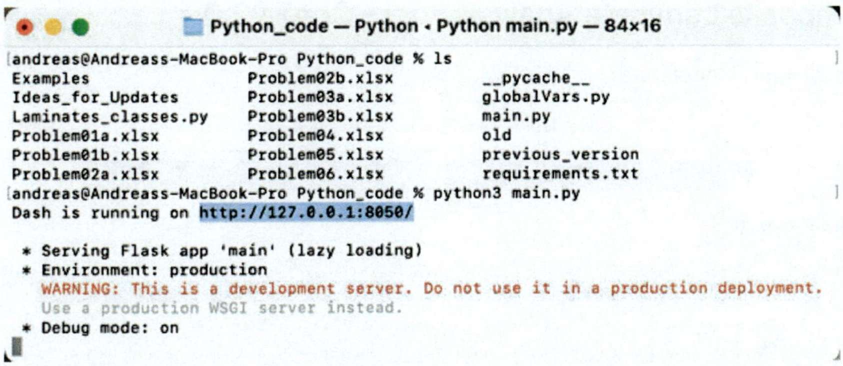

Fig. 3.1 Execution of the main Python file in a command-line interface

The following command, which must be executed in a command-line interface, should be executed once to install the current versions of the required libraries:

```
pip install -r requirements.txt
```

Then, the command

```
python3 main.py
```

starts the actual CLAT program. After executing this main Python file, the program provides in a shell a standard URL address, which can be copied to any common web browser such as Safari or Mozilla Firefox (see Fig. 3.1):

```
http://127.0.0.1:8050/
```

Copying the URL address to a web browser opens the CLAT graphical interface as indicated in Fig. 3.2.

The upper part of the interface (see `Import/Export Model Data`) allows to open and save model files. All the models are saved in a classical MS Excel table format. The `Model Properties` part of the interface allows defining the material and geometrical properties of each single lamina (layer) of a laminate. The used variables for elastic, strength, and geometrical properties are explained in detail in Table 3.1. This part of the interface also allows a simple addition or deletion of layers or a selective inactivation of certain layers. The `Load Matrix` part allows the definition of the length-normalized internal reactions, i.e., normal and shear forces and bending or twisting moments. The total thickness of the entire laminate is provided in the `Laminate Thickness` part.

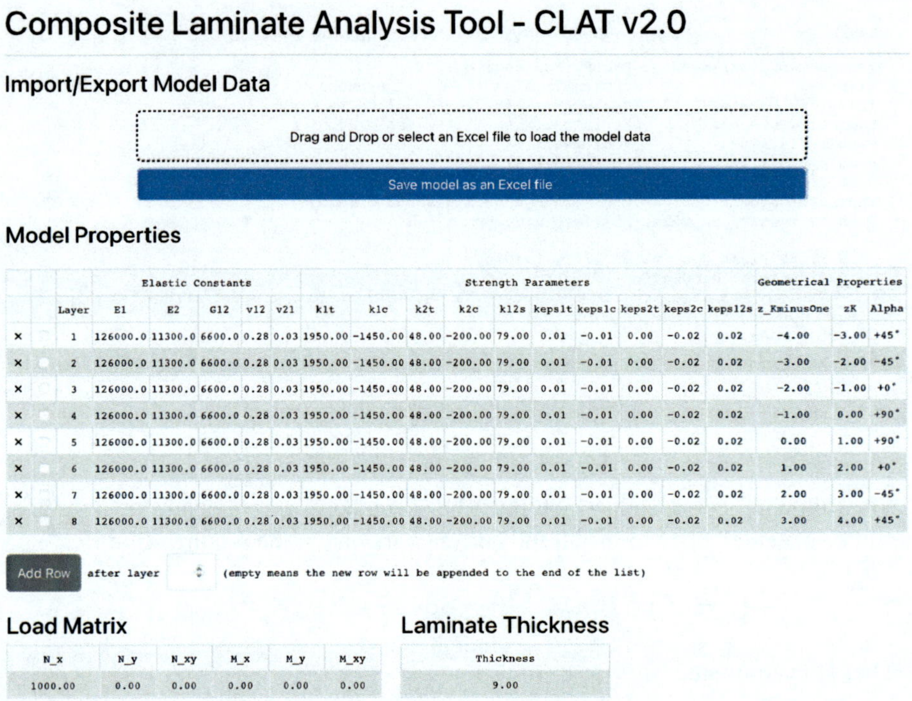

			Elastic Constants					Strength Parameters										Geometrical Properties		
		Layer	E1	E2	G12	v12	v21	k1t	k1c	k2t	k2c	k12s	keps1t	keps1c	keps2t	keps2c	keps12s	z_KminusOne	zK	Alpha
×		1	126000.0	11300.0	6600.0	0.28	0.03	1950.00	-1450.00	48.00	-200.00	79.00	0.01	-0.01	0.00	-0.02	0.02	-4.00	-3.00	+45°
×		2	126000.0	11300.0	6600.0	0.28	0.03	1950.00	-1450.00	48.00	-200.00	79.00	0.01	-0.01	0.00	-0.02	0.02	-3.00	-2.00	-45°
×		3	126000.0	11300.0	6600.0	0.28	0.03	1950.00	-1450.00	48.00	-200.00	79.00	0.01	-0.01	0.00	-0.02	0.02	-2.00	-1.00	+0°
×		4	126000.0	11300.0	6600.0	0.28	0.03	1950.00	-1450.00	48.00	-200.00	79.00	0.01	-0.01	0.00	-0.02	0.02	-1.00	0.00	+90°
×		5	126000.0	11300.0	6600.0	0.28	0.03	1950.00	-1450.00	48.00	-200.00	79.00	0.01	-0.01	0.00	-0.02	0.02	0.00	1.00	+90°
×		6	126000.0	11300.0	6600.0	0.28	0.03	1950.00	-1450.00	48.00	-200.00	79.00	0.01	-0.01	0.00	-0.02	0.02	1.00	2.00	+0°
×		7	126000.0	11300.0	6600.0	0.28	0.03	1950.00	-1450.00	48.00	-200.00	79.00	0.01	-0.01	0.00	-0.02	0.02	2.00	3.00	-45°
×		8	126000.0	11300.0	6600.0	0.28	0.03	1950.00	-1450.00	48.00	-200.00	79.00	0.01	-0.01	0.00	-0.02	0.02	3.00	4.00	+45°

N_x	N_y	N_xy	M_x	M_y	M_xy
1000.00	0.00	0.00	0.00	0.00	0.00

Thickness
9.00

Fig. 3.2 CLAT graphical interface: definition of model properties

The following options of the main CLAT interface (see Fig. 3.3, very upper part) allow to open saved models, operate the update mode as well as define the computational approach for particular questions[1]:

- `Existing models...`: It allows to open saved models in MS Excel format (see Figs. 3.4 and 3.5 for formatting details).
- `Regenerate graphs`: It allows to manually regenerate all the result graphs.
- ☑`Automatic Update`: This default option can be deactivated to reduce the processing time. The user may need to click `Regenerate graphs` to show updated results.
- ☐`Failure Analysis`: It allows to calculate the strength ratios R based on the maximum stress, maximum strain, Tsai-Hill, and Tsai-Wu criteria at the bottom, in the middle, and at the top of each lamina of a laminate. A strength ratio of $R \leq 1$ indicates the failure of a particular lamina.
- ☐`Ply-By-Ply Analysis`: It allows to calculate the successive failure of single lamina in a laminate based on the Tsai-Wu criterion and to repeat the calcu-

[1] A linear-elastic stress and strain analysis based on the classical laminate theory does not require the activation of a particular computation mode since it is the default approach and executed in CLAT without additional interaction.

Table 3.1 Variable names used in the CLAT interface for model definition (the provided units form a consistent set of units); see Fig. 3.2

Variable name	Explanation
Elastic constants	
`E1`	Longitudinal modulus, E_1 in MPa
`E2`	Transverse modulus, E_2 in MPa
`G12`	In-plane shear modulus, G_{12} in MPa
`v12`	Major Poisson's ratio, ν_{12} in –
`v21`	Minor Poisson's ratio, ν_{21} in –
Strength parameters	
`k1t`	Longitudinal tensile strength, k_{1t} in MPa
`k1c`	Longitudinal compressive strength, k_{1c} in MPa
`k2t`	Transverse tensile strength, k_{2t} in MPa
`k2c`	Transverse compressive strength, k_{2c} in MPa
`k12s`	In-plane shear strength, k_{12s} in MPa
`keps1t`	Longitudinal tensile failure strain, k^{ε}_{1t} in –
`keps1c`	Longitudinal compressive failure strain, k^{ε}_{1c} in –
`keps2t`	Transverse tensile failure strain, k^{ε}_{2t} in –
`keps2c`	Transverse compressive failure strain, k^{ε}_{2c} in –
`keps12s`	In-plane shear failure strain, k^{ε}_{12s} in –
Geometrical properties	
`z_KminusOne`	Lower z-coordinate of layer k (see Fig. 2.11), z_{k-1} in mm
`zK`	Upper z-coordinate of layer k (see Fig. 2.11), z_k in mm
`Alpha`	Rotation angle (see Fig. 1.2), α in degrees
`Thickness`	Thickness of the entire laminate (see Fig. 2.11), t in mm
Load parameters	
`N_x`	Normalized internal normal force in x-direction, N^{n}_x in N/mm
`N_y`	Normalized internal normal force in y-direction, N^{n}_y in N/mm
`N_xy`	Normalized internal shear force, N^{n}_{xy} in N/mm
`M_x`	Normalized internal bending moment (see Fig. 2.10), M^{n}_x in Nmm/mm
`M_y`	Normalized internal bending moment (see Fig. 2.10), M^{n}_y in Nmm/mm
`M_xy`	Normalized internal twisting moment (see Fig. 2.10), M^{n}_{xy} in Nmm/mm

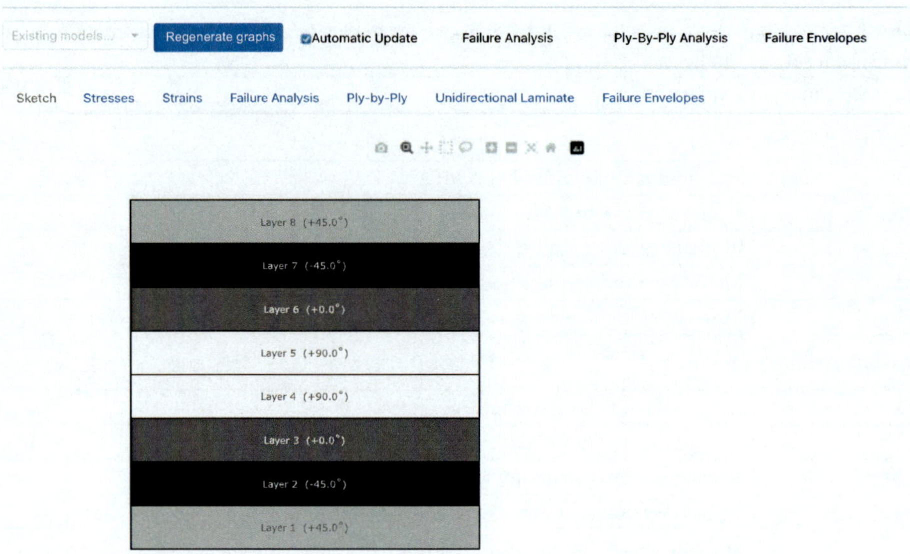

Fig. 3.3 CLAT graphical interface: analysis options and laminate representation

lation with a "reduced" laminate, i.e., a laminate where the failed lamina/laminae was/were deactivated.

- ☐`Failure Envelopes`: It allows to calculate for unidirectional (i.e., just one single layer) laminates the representations of failure envelopes, i.e., the maximum failure load in the form of Cartesian plots based on the angle of the UD fiber orientation. Considered are the maximum stress, maximum strain, Tsai-Hill, and Tsai-Wu criteria.

The following options of the main CLAT interface (see Fig. 3.3) allow to display the laminate composition and the calculation results:

- `Sketch`: This option shows the layer-wise stacking of the laminate by providing the layer numbers and the assigned rotational angles.
- `Stresses`: It shows graphs of the stress distributions over the laminate thickness in the local (*1–2*) as well as the global (x–y) coordinate systems. In addition, a numerical output in table format can be displayed.
- `Strains`: It shows graphs of the strain distributions over the laminate thickness in the local (*1–2*) as well as the global (x–y) coordinate systems. In addition, a numerical output in table format can be displayed.
- `Failure Analysis`: It displays a table of the strength ratio R at the bottom, in the middle, and at the top of each layer based on the maximum stress, maximum strain, Tsai-Hill, and Tsai-Wu criteria. It requires setting the ☑`Failure Analysis` option.
- `Ply-by-Ply`: It displays the stress and strain distributions in the x-direction for different steps of the ply-by-ply failure analysis, as well as the macroscopic stress-

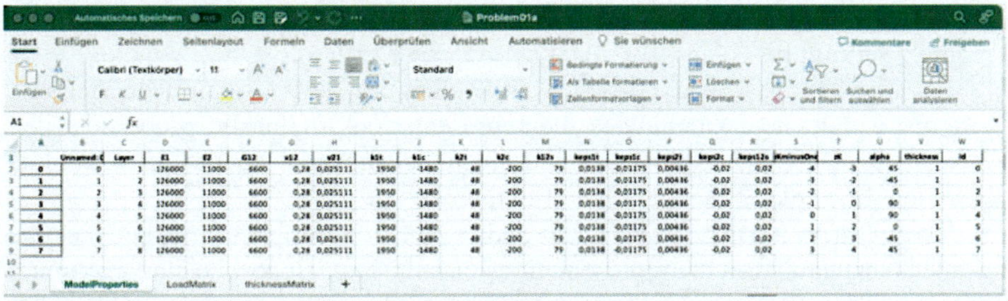

Fig. 3.4 Model definition in Microsoft Excel: material and geometrical properties of each layer (lamina)

Fig. 3.5 Model definition in Microsoft Excel: **a** load matrix and **b** laminate thickness

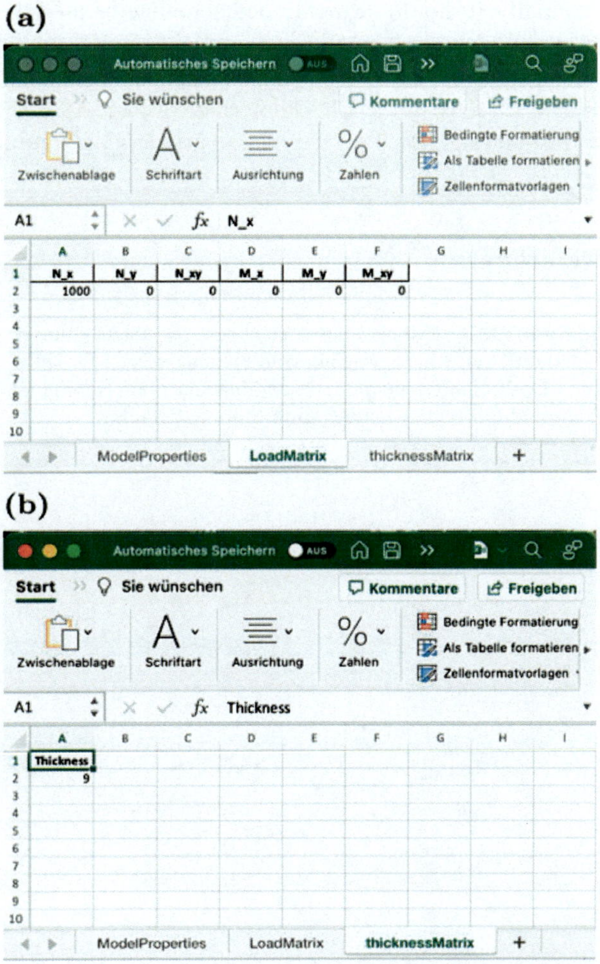

strain diagram under consideration of the layer-wise failure of single laminae. It requires setting the ☑`Ply-By-Ply Analysis` option.

- `Unidirectional Laminate`: This result option plots for unidirectional (i.e., just one single layer) laminates the representations of elastic properties, i.e., E_x/E_1, E_y/E_1, G_{xy}/G_{12}, ν_{xy}, η_x, and η_y, in the form of polar or Cartesian plots based on the angle of the UD fiber orientation.
- `Failure Envelopes`: This result option plots for unidirectional (i.e., just one single layer) laminates the representations of failure envelopes, i.e., the maximum failure load in the form of Cartesian plots based on the angle of the UD fiber orientation. Considered are the maximum stress, maximum strain, Tsai-Hill, and Tsai-Wu criteria. It requires to set the ☑`Failure Envelopes` option.

Finally, it should be mentioned here that the model definition or modification can be simply done in the MS Excel model files. Figure 3.4 shows the defined model of Fig. 3.2 as a classical MS Excel table. Thus, the user can decide to create and modify models directly in the CLAT interface or in a MS Excel table. Figure 3.5 shows the corresponding tables for the load matrix and the laminate thickness.

References

Bernard J (2016) Python recipes handbook: a problem-solution approach. Apress, Berkeley
Kalb I (2016) Learn to program with Python. Apress, Berkeley
Lee KD (2011) Python programming fundamentals. Springer, London
Pilgrim M (2004) Dive into Python. Apress, Berkeley

Chapter 4
Application Examples

Abstract This chapter presents the solution of some standard calculation problems based on the classical laminate theory. Each solution presents the major result plots or tables obtained in CLAT. The presented problems comprise the calculation of stresses and strains in symmetric and asymmetric laminates, the application of failure criteria, and the ply-by-ply failure of laminates. In addition, the characterization of basic properties of unidirectional laminae based on pole diagrams for the elastic properties and failure envelopes is presented.

4.1 Introduction

The following sections will cover some standard calculation problems which are solved using CLAT. The solution procedure without a computational tool is presented in Öchsner (2021). All the presented problems are based on the same AS4 carbon/3501-6 epoxy prepreg[1] system. The effective lamina properties can be taken from Table 4.1. Other sets of complete mechanical properties for unidirectional laminae can be found, for example, in Soden et al. (1998), Kaw (2006).

The theoretical prediction of lamina properties, i.e., the so-called micromechanics, is not covered in CLAT, and the interested reader is referred to the specialized literature, e.g., Herakovich (2012), Buryachenko (2007), and Dvorak (2013).

[1] A prepreg is a ready to mold or cure sheet which may be tow, tape, cloth, or mat impregnated with resin MIL-HDBK-17/1F (2002).

A. Öchsner and R. Makvandi, *A Numerical Approach to the Classical Laminate Theory of Composite Materials*, Advanced Structured Materials 189,
https://doi.org/10.1007/978-3-031-32975-3_4

Table 4.1 Selected mechanical properties of a unidirectional lamina (fiber: AS4 carbon; matrix: 3501-6 epoxy; fiber volume fraction of the prepreg: 0.6). Adapted from Soden et al. (1998)

Property	Value
Elastic constants	
Longitudinal modulus, E_1 in MPa	126000
Transverse modulus, E_2 in MPa	11000
In-plane shear modulus, G_{12} in MPa	6600
Major Poisson's ratio, ν_{12} in –	0.28
Strength parameters	
Longitudinal tensile strength, k_{1t} in MPa	1950
Longitudinal compressive strength, k_{1c} in MPa	−1480
Transverse tensile strength, k_{2t} in MPa	48
Transverse compressive strength, k_{2c} in MPa	−200
In-plane shear strength, k_{12s} in MPa	79
Longitudinal tensile failure strain, k^{ε}_{1t} in %	1.38
Longitudinal compressive failure strain, k^{ε}_{1c} in %	−1.175
Transverse tensile failure strain, k^{ε}_{2t} in %	0.436
Transverse compressive failure strain, k^{ε}_{2c} in %	−2.0
In-plane shear failure strain, k^{ε}_{12s} in %	2

4.2 Problem 1: Stresses and Strains in a Symmetric Laminate

Given is a symmetric 8-layer laminate of thickness $t = 8$ mm as shown in Fig. 4.1. Each lamina has the same thickness ($t/8$) and identical material properties, which can be taken from Table 4.1. However, the orientation of each lamina changes from layer to layer as indicated in Fig. 4.1.

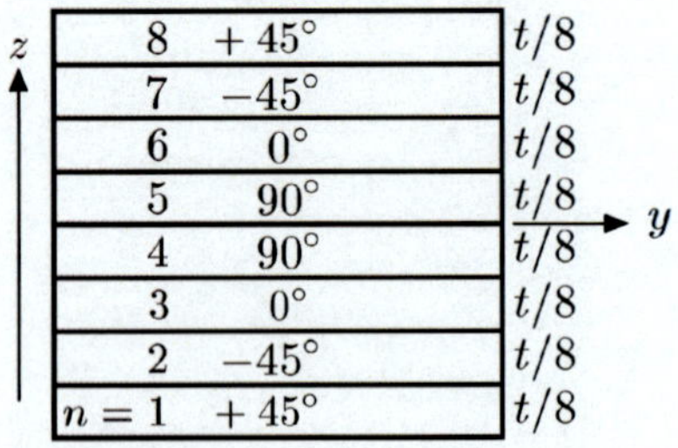

Fig. 4.1 Symmetric laminate: $[\pm 45°/0°/90°]_s$

Consider (a) a pure tensile load case with $N_x^{\mathrm{n}} = 1000$ N/mm (all other forces and moments are equal to zero) and (b) a pure bending load case with $M_x^{\mathrm{n}} = 1000$ Nmm/mm (all other moments and forces are equal to zero) to determine the stresses and strains in each lamina in the global x–y coordinate system as well as in the principal material coordinates (*1–2* coordinate systems) of each lamina.

- Solution for the laminate loaded by $N_x^{\mathrm{n}} = 1000$ N/mm

The model definition of the symmetric laminate $[\pm 45°/0°/90°]_s$ in CLAT is shown in Fig. 4.2. It can be seen that every single layer is represented with its material and geometrical properties by a single row. The applied load, i.e., $N_x^{\mathrm{n}} = 1000$ N/mm, is represented in the `Load Matrix` field, and the thickness of the entire laminate ($t = 8$ mm) is shown in the `Laminate Thickness` field.

The graphical representation of the $[\pm 45°/0°/90°]_s$ laminate in CLAT is given in Fig. 4.3. Layer numbers and angles of rotation are provided, and each angle is represented by a different color.

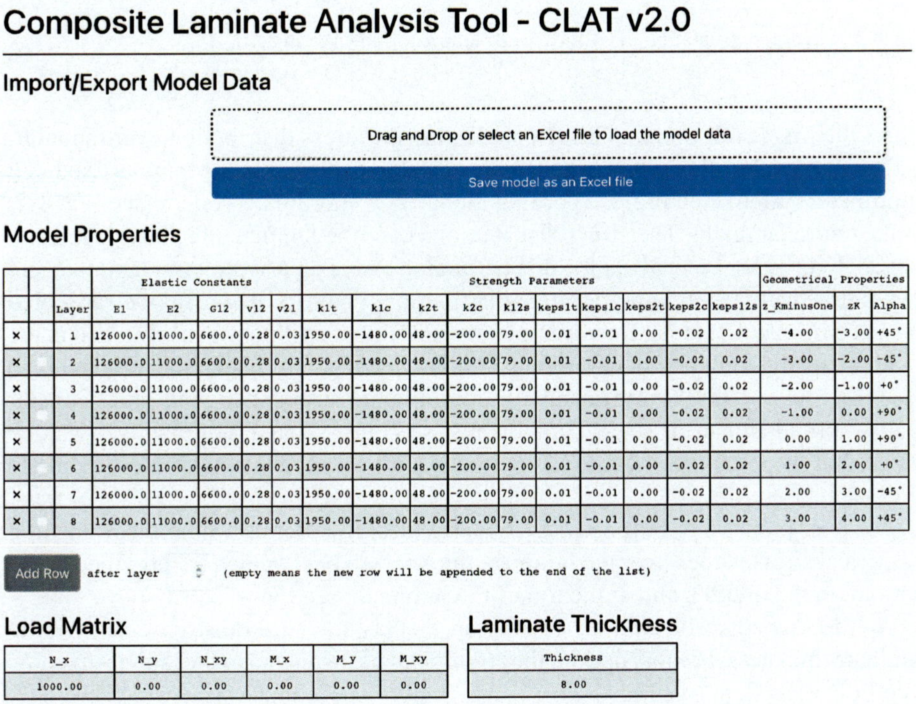

Composite Laminate Analysis Tool - CLAT v2.0

Import/Export Model Data

Drag and Drop or select an Excel file to load the model data

Save model as an Excel file

Model Properties

			Elastic Constants					Strength Parameters										Geometrical Properties		
		Layer	E1	E2	G12	v12	v21	k1t	k1c	k2t	k2c	k12s	keps1t	keps1c	keps2t	keps2c	keps12s	z_KminusOne	zK	Alpha
×		1	126000.0	11000.0	6600.0	0.28	0.03	1950.00	-1480.00	48.00	-200.00	79.00	0.01	-0.01	0.00	-0.02	0.02	-4.00	-3.00	+45°
×		2	126000.0	11000.0	6600.0	0.28	0.03	1950.00	-1480.00	48.00	-200.00	79.00	0.01	-0.01	0.00	-0.02	0.02	-3.00	-2.00	-45°
×		3	126000.0	11000.0	6600.0	0.28	0.03	1950.00	-1480.00	48.00	-200.00	79.00	0.01	-0.01	0.00	-0.02	0.02	-2.00	-1.00	+0°
×		4	126000.0	11000.0	6600.0	0.28	0.03	1950.00	-1480.00	48.00	-200.00	79.00	0.01	-0.01	0.00	-0.02	0.02	-1.00	0.00	+90°
×		5	126000.0	11000.0	6600.0	0.28	0.03	1950.00	-1480.00	48.00	-200.00	79.00	0.01	-0.01	0.00	-0.02	0.02	0.00	1.00	+90°
×		6	126000.0	11000.0	6600.0	0.28	0.03	1950.00	-1480.00	48.00	-200.00	79.00	0.01	-0.01	0.00	-0.02	0.02	1.00	2.00	+0°
×		7	126000.0	11000.0	6600.0	0.28	0.03	1950.00	-1480.00	48.00	-200.00	79.00	0.01	-0.01	0.00	-0.02	0.02	2.00	3.00	-45°
×		8	126000.0	11000.0	6600.0	0.28	0.03	1950.00	-1480.00	48.00	-200.00	79.00	0.01	-0.01	0.00	-0.02	0.02	3.00	4.00	+45°

Add Row after layer (empty means the new row will be appended to the end of the list)

Load Matrix

N_x	N_y	N_xy	M_x	M_y	M_xy
1000.00	0.00	0.00	0.00	0.00	0.00

Laminate Thickness

Thickness
8.00

Fig. 4.2 Model definition in CLAT for the laminate shown in Fig. 4.1, tensile load case

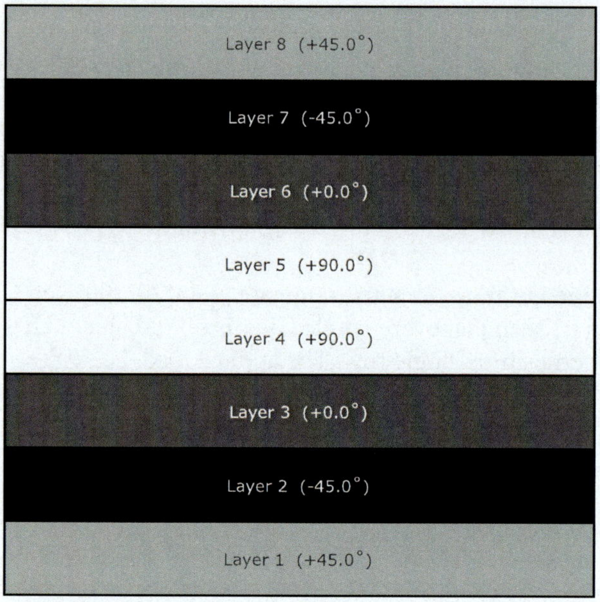

Fig. 4.3 Layer representation in CLAT for the laminate shown in Fig. 4.1

As the first result option, CLAT can display the stress distribution over the laminate thickness. For this option, the stresses can be displayed in the lamina-fixed *1–2* coordinate system (see Fig. 2.2) or in a global x–y coordinate system (see Fig. 1.2) of the entire laminate. The stress distributions over the laminate thickness shown in Figs. 4.4, 4.5, and 4.6 provide for this particular case, i.e., a symmetric laminate and symmetric load case, symmetric stress distributions in the local (*1–2*) as well as global (x–y) coordinate systems. However, clearly different distributions are obtained in the local compared to the global system. It can be seen in addition that the uniaxial load case (i.e., $N_x^{\text{n}} = 1000\,\text{N/mm}$) results in a multiaxial stress state with other considerable normal and shear stress components; see Figs. 4.5 and 4.6. Finally, it can be stated that the considered case results in constant stress values within every single lamina.

For post-processing purposes, CLAT provides, in addition to the diagrams, the numerical stress values over the laminate thickness. These values are provided at the bottom, in the middle, and at the top of the laminate; see Fig. 4.7.

As the second result option, CLAT can display the strain distribution over the laminate thickness. For this option, the strains can be displayed in the lamina-fixed *1–2* coordinate system (see Fig. 2.2) or in a global x–y coordinate system (see Fig. 1.2) of the entire laminate. The strain distributions over the laminate thickness shown in Figs. 4.8, 4.9, and 4.10 provide for this particular case, i.e., a symmetric laminate

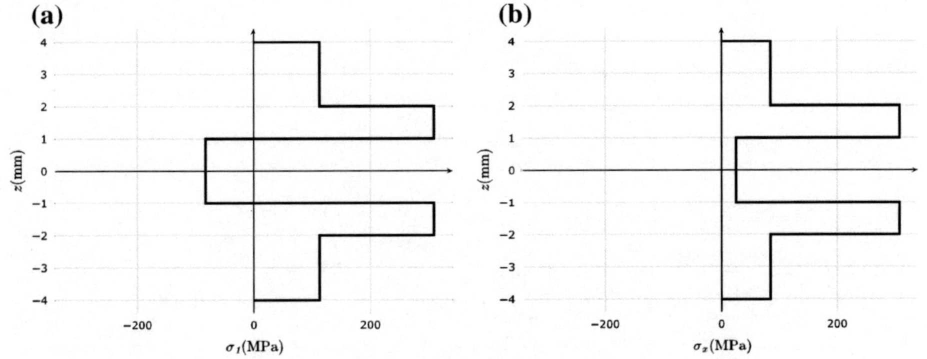

Fig. 4.4 Distribution of selected field quantities over the laminate thickness for the tensile load case: **a** normal stress $\sigma_1(z)$ and **b** normal stress $\sigma_x(z)$

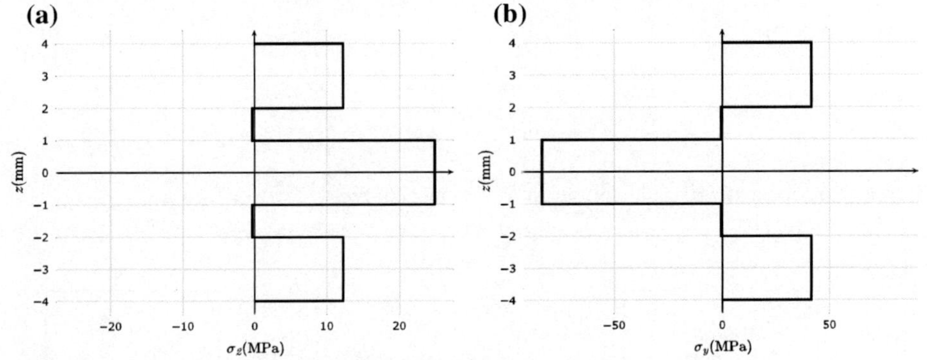

Fig. 4.5 Distribution of selected field quantities over the laminate thickness for the tensile load case: **a** normal stress $\sigma_2(z)$ and **b** normal stress $\sigma_y(z)$

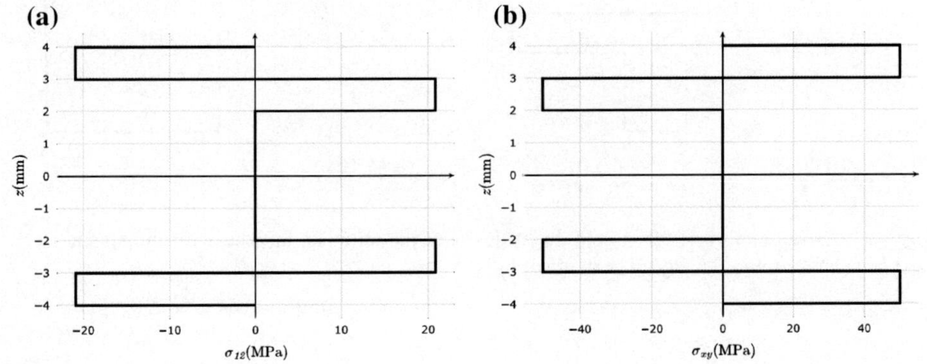

Fig. 4.6 Distribution of selected field quantities over the laminate thickness for the tensile load case: **a** shear stress $\sigma_{12}(z)$ and **b** shear stress $\sigma_{xy}(z)$

Stress Table

Export

Layer no.	Position	sigma_1	sigma_2	sigma_12	sigma_x	sigma_y	sigma_xy
1	Bottom	112.713	12.287	-20.875	83.375	41.625	50.213
1	Center	112.713	12.287	-20.875	83.375	41.625	50.213
1	Top	112.713	12.287	-20.875	83.375	41.625	50.213
2	Bottom	112.713	12.287	20.875	83.375	41.625	-50.213
2	Center	112.713	12.287	20.875	83.375	41.625	-50.213
2	Top	112.713	12.287	20.875	83.375	41.625	-50.213
3	Bottom	308.349	-0.327	0.000	308.349	-0.327	0.000
3	Center	308.349	-0.327	0.000	308.349	-0.327	0.000
3	Top	308.349	-0.327	0.000	308.349	-0.327	0.000
4	Bottom	-82.922	24.901	0.000	24.901	-82.922	0.000
4	Center	-82.922	24.901	0.000	24.901	-82.922	0.000
4	Top	-82.922	24.901	0.000	24.901	-82.922	0.000
5	Bottom	-82.922	24.901	0.000	24.901	-82.922	0.000
5	Center	-82.922	24.901	0.000	24.901	-82.922	0.000
5	Top	-82.922	24.901	0.000	24.901	-82.922	0.000
6	Bottom	308.349	-0.327	0.000	308.349	-0.327	0.000
6	Center	308.349	-0.327	0.000	308.349	-0.327	0.000
6	Top	308.349	-0.327	0.000	308.349	-0.327	0.000
7	Bottom	112.713	12.287	20.875	83.375	41.625	-50.213
7	Center	112.713	12.287	20.875	83.375	41.625	-50.213
7	Top	112.713	12.287	20.875	83.375	41.625	-50.213
8	Bottom	112.713	12.287	-20.875	83.375	41.625	50.213
8	Center	112.713	12.287	-20.875	83.375	41.625	50.213
8	Top	112.713	12.287	-20.875	83.375	41.625	50.213

Fig. 4.7 Numerical stress results for the laminate shown in Fig. 4.1, tensile load case

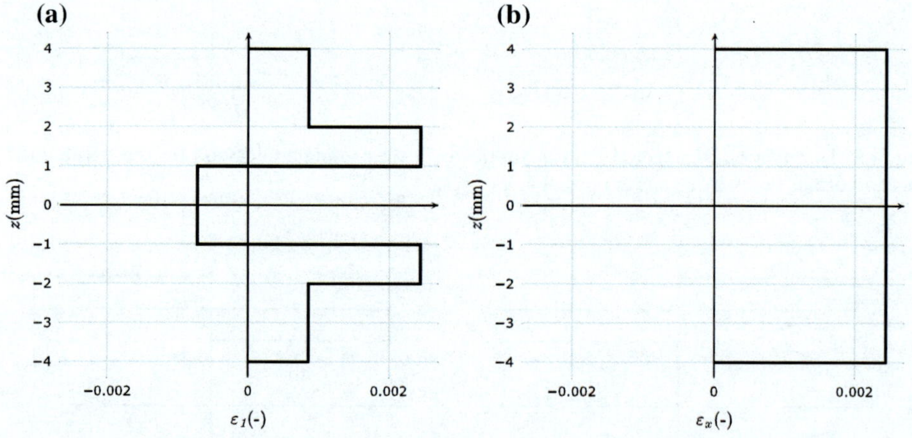

Fig. 4.8 Distribution of selected field quantities over the laminate thickness for the tensile load case: **a** normal strain $\epsilon_1(z)$ and **b** normal strain $\epsilon_x(z)$

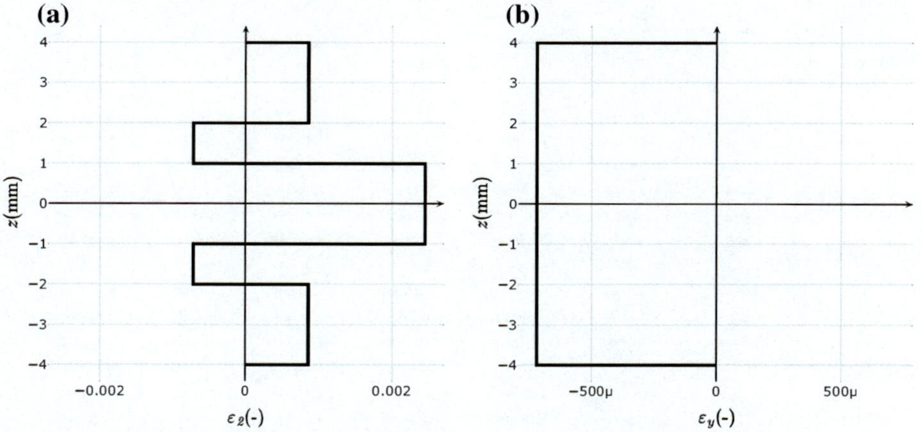

Fig. 4.9 Distribution of selected field quantities over the laminate thickness for the tensile load case: **a** normal strain $\epsilon_2(z)$ and **b** normal strain $\epsilon_y(z)$

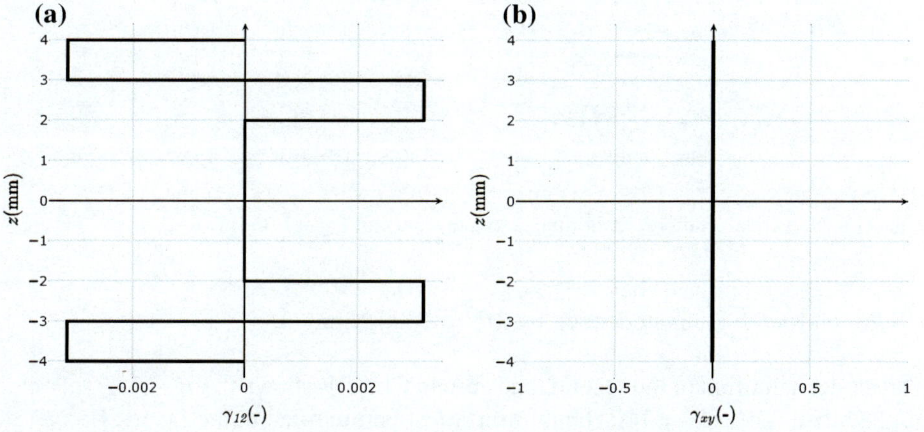

Fig. 4.10 Distribution of selected field quantities over the laminate thickness for the tensile load case: **a** shear strain $\gamma_{12}(z)$ and **b** shear strain $\gamma_{xy}(z)$

and symmetric load case, symmetric strain distributions in the local (*1–2*) as well as global (x–y) coordinate systems. However, the global coordinate representations provide a single constant value across the entire laminate thickness.

As in the case of the stress distribution, numerical values for the strains can be obtained at the bottom, in the middle, and at the top of each lamina; see Fig. 4.11.

Strain Table

Export

Layer no.	Position	eps_1	eps_2	gamma_12	eps_x	eps_y	gamma_xy
1	Bottom	0.000867	0.000867	-0.003163	0.002448	-0.000715	0.000000
1	Center	0.000867	0.000867	-0.003163	0.002448	-0.000715	0.000000
1	Top	0.000867	0.000867	-0.003163	0.002448	-0.000715	0.000000
2	Bottom	0.000867	0.000867	0.003163	0.002448	-0.000715	0.000000
2	Center	0.000867	0.000867	0.003163	0.002448	-0.000715	0.000000
2	Top	0.000867	0.000867	0.003163	0.002448	-0.000715	0.000000
3	Bottom	0.002448	-0.000715	0.000000	0.002448	-0.000715	0.000000
3	Center	0.002448	-0.000715	0.000000	0.002448	-0.000715	0.000000
3	Top	0.002448	-0.000715	0.000000	0.002448	-0.000715	0.000000
4	Bottom	-0.000715	0.002448	0.000000	0.002448	-0.000715	0.000000
4	Center	-0.000715	0.002448	0.000000	0.002448	-0.000715	0.000000
4	Top	-0.000715	0.002448	0.000000	0.002448	-0.000715	0.000000
5	Bottom	-0.000715	0.002448	0.000000	0.002448	-0.000715	0.000000
5	Center	-0.000715	0.002448	0.000000	0.002448	-0.000715	0.000000
5	Top	-0.000715	0.002448	0.000000	0.002448	-0.000715	0.000000
6	Bottom	0.002448	-0.000715	0.000000	0.002448	-0.000715	0.000000
6	Center	0.002448	-0.000715	0.000000	0.002448	-0.000715	0.000000
6	Top	0.002448	-0.000715	0.000000	0.002448	-0.000715	0.000000
7	Bottom	0.000867	0.000867	0.003163	0.002448	-0.000715	0.000000
7	Center	0.000867	0.000867	0.003163	0.002448	-0.000715	0.000000
7	Top	0.000867	0.000867	0.003163	0.002448	-0.000715	0.000000
8	Bottom	0.000867	0.000867	-0.003163	0.002448	-0.000715	0.000000
8	Center	0.000867	0.000867	-0.003163	0.002448	-0.000715	0.000000
8	Top	0.000867	0.000867	-0.003163	0.002448	-0.000715	0.000000

Fig. 4.11 Numerical strain results for the laminate shown in Fig. 4.1, tensile load case

- Solution for the laminate loaded by $M_x^{\text{n}} = 1000$ Nmm/mm

The model definition of the second load case in CLAT is shown in Fig. 4.12. Only the applied load (i.e., $M_x^{\text{n}} = 1000\,\text{Nmm/mm}$) must be modified in the `Load Matrix` field, and all other material and geometrical properties remain unchanged.

The stress distributions over the laminate thickness are shown in Figs. 4.13, 4.14, and 4.15 in the lamina-fixed *1–2* coordinate system (see Fig. 2.2) or in a global *x–y* coordinate system (see Fig. 1.2) of the entire laminate. It can be seen that for this particular case, i.e., a symmetric laminate and bending load case, anti-symmetric stress distributions are obtained in both coordinate systems.

For post-processing purposes, the numerical stress values over the laminate thickness, i.e., at the bottom, in the middle, and at the top of the laminate, are provided in Fig. 4.16.

The strains distributions over the laminate thickness are shown in Figs. 4.17, 4.18, and 4.19 in the lamina-fixed *1–2* coordinate system (see Fig. 2.2) or in a global *x–y* coordinate system (see Fig. 1.2) of the entire laminate. It can be seen that for this particular case, i.e., a symmetric laminate and bending load case, complex anti-

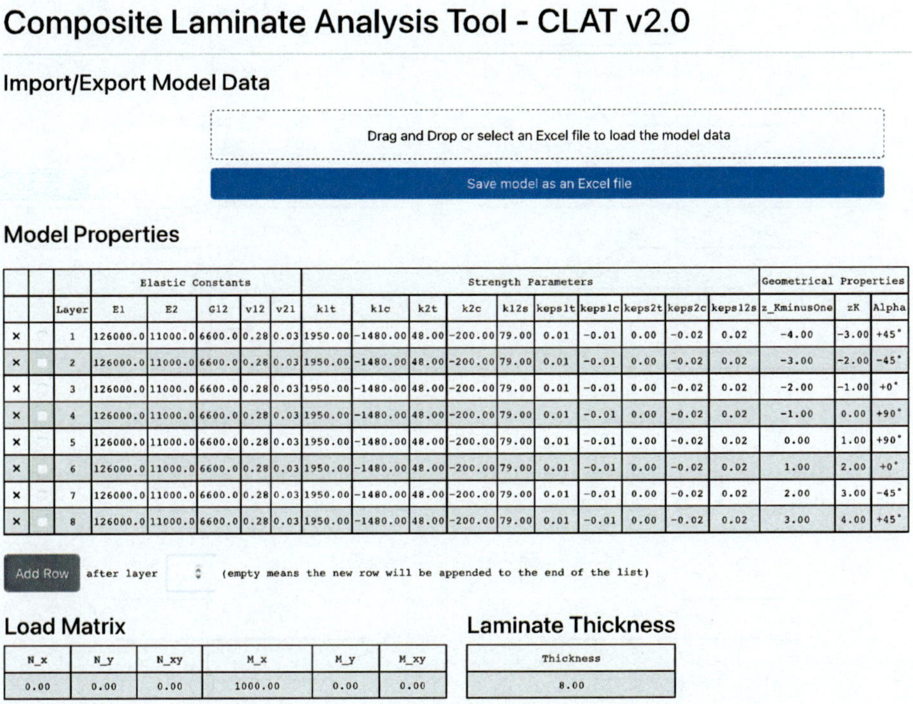

Composite Laminate Analysis Tool - CLAT v2.0

Import/Export Model Data

Drag and Drop or select an Excel file to load the model data

Save model as an Excel file

Model Properties

			Elastic Constants					Strength Parameters										Geometrical Properties		
		Layer	E1	E2	G12	v12	v21	k1t	k1c	k2t	k2c	k12s	keps1t	keps1c	keps2t	keps2c	keps12s	z_KminusOne	zK	Alpha
×		1	126000.0	11000.0	6600.0	0.28	0.03	1950.00	-1480.00	48.00	-200.00	79.00	0.01	-0.01	0.00	-0.02	0.02	-4.00	-3.00	+45°
×		2	126000.0	11000.0	6600.0	0.28	0.03	1950.00	-1480.00	48.00	-200.00	79.00	0.01	-0.01	0.00	-0.02	0.02	-3.00	-2.00	-45°
×		3	126000.0	11000.0	6600.0	0.28	0.03	1950.00	-1480.00	48.00	-200.00	79.00	0.01	-0.01	0.00	-0.02	0.02	-2.00	-1.00	+0°
×		4	126000.0	11000.0	6600.0	0.28	0.03	1950.00	-1480.00	48.00	-200.00	79.00	0.01	-0.01	0.00	-0.02	0.02	-1.00	0.00	+90°
×		5	126000.0	11000.0	6600.0	0.28	0.03	1950.00	-1480.00	48.00	-200.00	79.00	0.01	-0.01	0.00	-0.02	0.02	0.00	1.00	+90°
×		6	126000.0	11000.0	6600.0	0.28	0.03	1950.00	-1480.00	48.00	-200.00	79.00	0.01	-0.01	0.00	-0.02	0.02	1.00	2.00	+0°
×		7	126000.0	11000.0	6600.0	0.28	0.03	1950.00	-1480.00	48.00	-200.00	79.00	0.01	-0.01	0.00	-0.02	0.02	2.00	3.00	-45°
×		8	126000.0	11000.0	6600.0	0.28	0.03	1950.00	-1480.00	48.00	-200.00	79.00	0.01	-0.01	0.00	-0.02	0.02	3.00	4.00	+45°

Add Row after layer (empty means the new row will be appended to the end of the list)

Load Matrix

N_x	N_y	N_xy	M_x	M_y	M_xy
0.00	0.00	0.00	1000.00	0.00	0.00

Laminate Thickness

Thickness
8.00

Fig. 4.12 Model definition in CLAT for the laminate shown in Fig. 4.1, bending load case

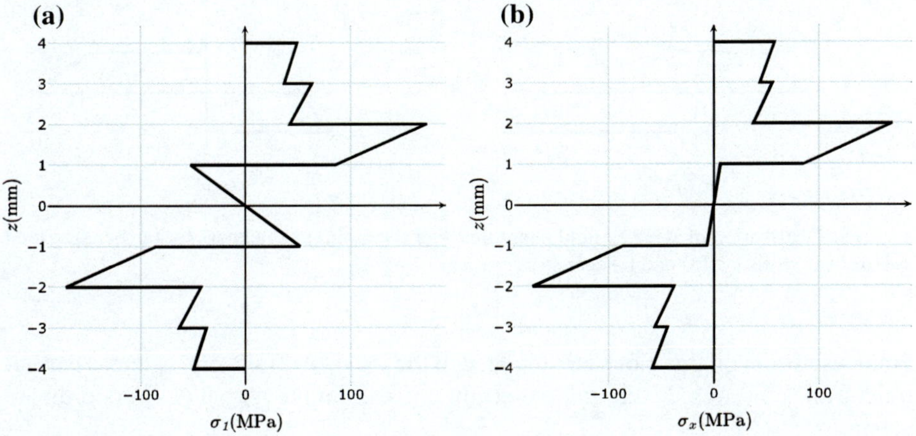

Fig. 4.13 Distribution of selected field quantities over the laminate thickness for the bending load case: **a** normal stress $\sigma_1(z)$ and **b** normal stress $\sigma_x(z)$

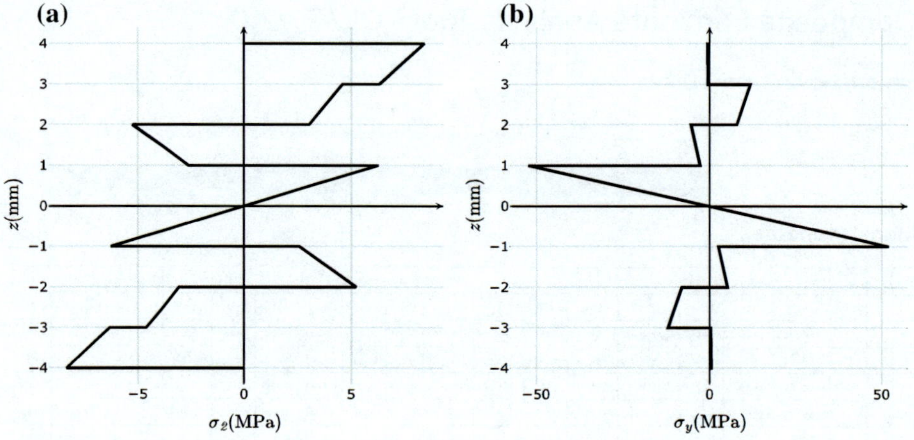

Fig. 4.14 Distribution of selected field quantities over the laminate thickness for the bending load case: **a** normal stress $\sigma_2(z)$ and **b** normal stress $\sigma_y(z)$

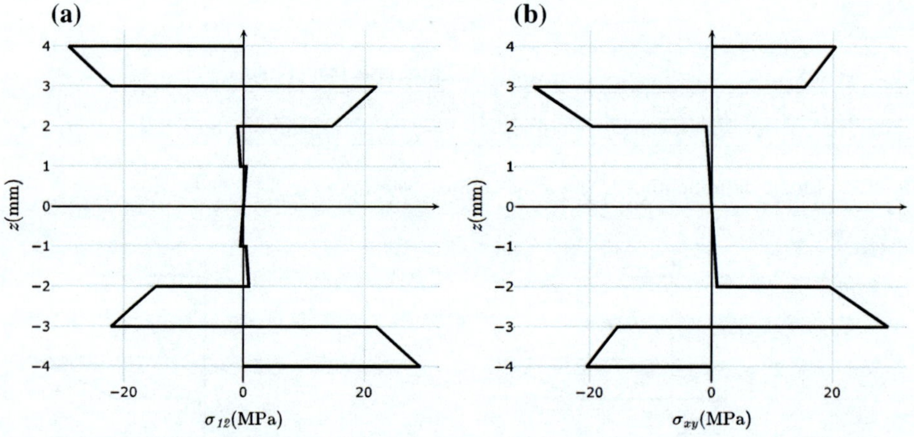

Fig. 4.15 Distribution of selected field quantities over the laminate thickness for the bending load case: **a** shear stress $\sigma_{12}(z)$ and **b** shear stress $\sigma_{xy}(z)$

symmetric strain distributions are obtained in the local (*1–2*) system, whereas perfect straight lines through the coordinate origin are seen in the global (x–y) coordinate system.

For post-processing purposes, the numerical strain values over the laminate thickness, i.e., at the bottom, in the middle, and at the top of the laminate, are provided in Fig. 4.20.

Stress Table

Export

Layer no.	Position	sigma_1	sigma_2	sigma_12	sigma_x	sigma_y	sigma_xy
1	Bottom	-49.389	-8.419	29.417	-58.321	0.513	-20.485
1	Center	-43.216	-7.366	25.740	-51.031	0.449	-17.925
1	Top	-37.042	-6.314	22.063	-43.741	0.385	-15.364
2	Bottom	-63.282	-4.622	-22.063	-56.015	-11.889	29.330
2	Center	-52.735	-3.852	-18.386	-46.679	-9.908	24.442
2	Top	-42.188	-3.081	-14.709	-37.343	-7.926	19.553
3	Bottom	-171.286	5.242	0.933	-171.286	5.242	0.933
3	Center	-128.464	3.932	0.700	-128.464	3.932	0.700
3	Top	-85.643	2.621	0.467	-85.643	2.621	0.467
4	Bottom	52.202	-6.267	-0.467	-6.267	52.202	0.467
4	Center	26.101	-3.133	-0.233	-3.133	26.101	0.233
4	Top	0.000	0.000	0.000	0.000	0.000	0.000
5	Bottom	0.000	0.000	0.000	0.000	0.000	0.000
5	Center	-26.101	3.133	0.233	3.133	-26.101	-0.233
5	Top	-52.202	6.267	0.467	6.267	-52.202	-0.467
6	Bottom	85.643	-2.621	-0.467	85.643	-2.621	-0.467
6	Center	128.464	-3.932	-0.700	128.464	-3.932	-0.700
6	Top	171.286	-5.242	-0.933	171.286	-5.242	-0.933
7	Bottom	42.188	3.081	14.709	37.343	7.926	-19.553
7	Center	52.735	3.852	18.386	46.679	9.908	-24.442
7	Top	63.282	4.622	22.063	56.015	11.889	-29.330
8	Bottom	37.042	6.314	-22.063	43.741	-0.385	15.364
8	Center	43.216	7.366	-25.740	51.031	-0.449	17.925
8	Top	49.389	8.419	-29.417	58.321	-0.513	20.485

Fig. 4.16 Numerical stress results for the laminate shown in Fig. 4.1, bending load case

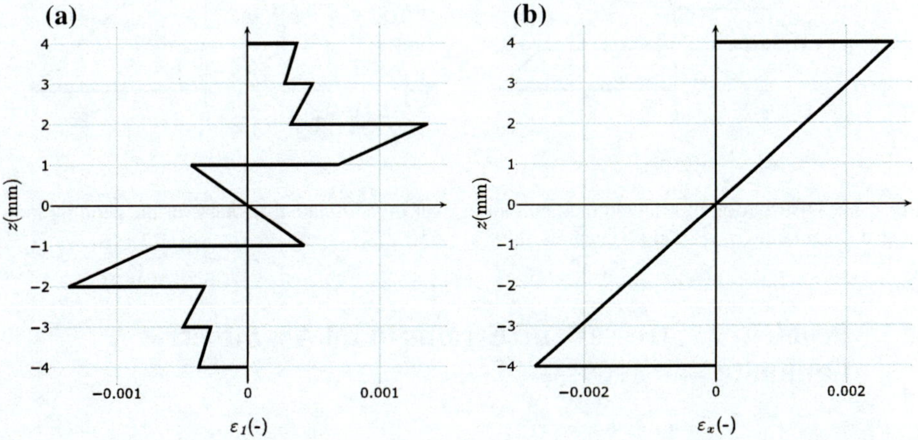

Fig. 4.17 Distribution of selected field quantities over the laminate thickness for the bending load case: **a** normal strain $\epsilon_I(z)$ and **b** normal strain $\epsilon_x(z)$

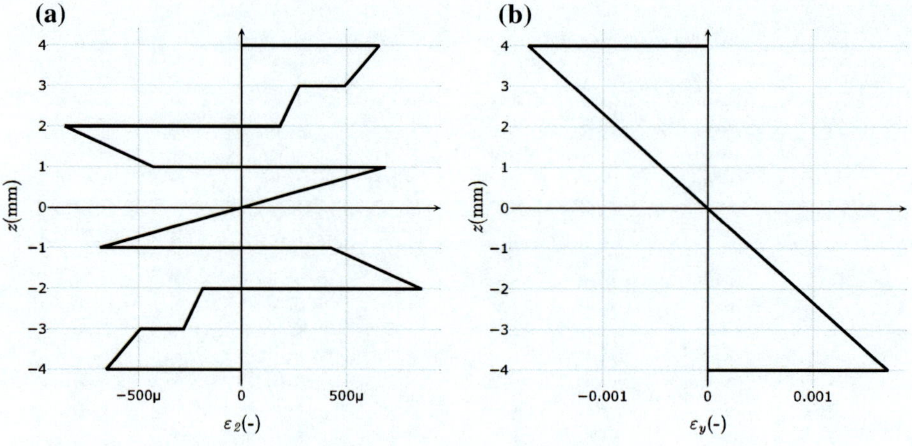

Fig. 4.18 Distribution of selected field quantities over the laminate thickness for the bending load case: **a** normal strain $\epsilon_2(z)$ and **b** normal strain $\epsilon_y(z)$

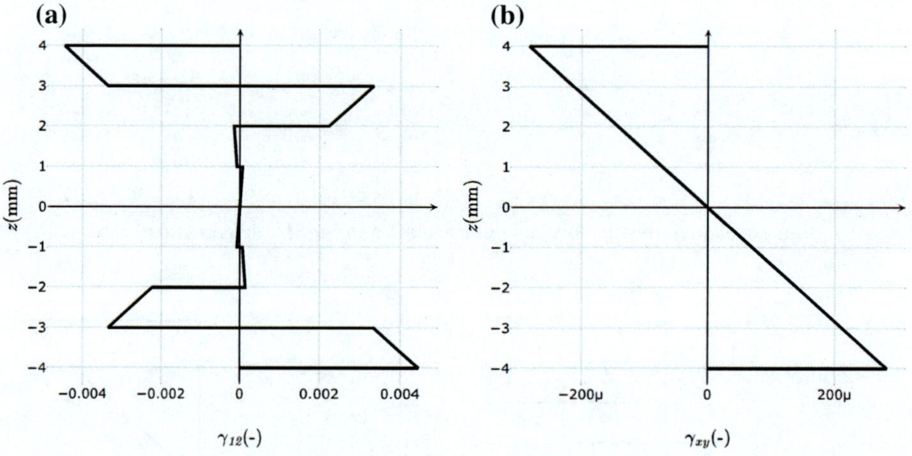

Fig. 4.19 Distribution of selected field quantities over the laminate thickness for the bending load case: **a** shear strain $\gamma_{12}(z)$ and **b** shear strain $\gamma_{xy}(z)$

4.3 Problem 2: Stresses and Strains in an Asymmetric Laminate

Given is an asymmetric 8-layer laminate of thickness $t = 8$ mm as shown in Fig. 4.21. Each lamina has the same thickness $(t/8)$ and identical material properties, which can be taken from Table 4.1. However, the orientation of each lamina changes from layer to layer as indicated in Fig. 4.21.

Strain Table

Export

Layer no.	Position	eps_1	eps_2	gamma_12	eps_x	eps_y	gamma_xy
1	Bottom	-0.000373	-0.000656	0.004457	-0.002743	0.001714	0.000283
1	Center	-0.000326	-0.000574	0.003900	-0.002400	0.001500	0.000247
1	Top	-0.000280	-0.000492	0.003343	-0.002057	0.001286	0.000212
2	Bottom	-0.000492	-0.000280	-0.003343	-0.002057	0.001286	0.000212
2	Center	-0.000410	-0.000233	-0.002786	-0.001714	0.001072	0.000177
2	Top	-0.000328	-0.000186	-0.002229	-0.001371	0.000857	0.000141
3	Bottom	-0.001371	0.000857	0.000141	-0.001371	0.000857	0.000141
3	Center	-0.001029	0.000643	0.000106	-0.001029	0.000643	0.000106
3	Top	-0.000686	0.000429	0.000071	-0.000686	0.000429	0.000071
4	Bottom	0.000429	-0.000686	-0.000071	-0.000686	0.000429	0.000071
4	Center	0.000214	-0.000343	-0.000035	-0.000343	0.000214	0.000035
4	Top	0.000000	0.000000	0.000000	0.000000	0.000000	0.000000
5	Bottom	0.000000	0.000000	0.000000	0.000000	0.000000	0.000000
5	Center	-0.000214	0.000343	0.000035	0.000343	-0.000214	-0.000035
5	Top	-0.000429	0.000686	0.000071	0.000686	-0.000429	-0.000071
6	Bottom	0.000686	-0.000429	-0.000071	0.000686	-0.000429	-0.000071
6	Center	0.001029	-0.000643	-0.000106	0.001029	-0.000643	-0.000106
6	Top	0.001371	-0.000857	-0.000141	0.001371	-0.000857	-0.000141
7	Bottom	0.000328	0.000186	0.002229	0.001371	-0.000857	-0.000141
7	Center	0.000410	0.000233	0.002786	0.001714	-0.001072	-0.000177
7	Top	0.000492	0.000280	0.003343	0.002057	-0.001286	-0.000212
8	Bottom	0.000280	0.000492	-0.003343	0.002057	-0.001286	-0.000212
8	Center	0.000326	0.000574	-0.003900	0.002400	-0.001500	-0.000247
8	Top	0.000373	0.000656	-0.004457	0.002743	-0.001714	-0.000283

Fig. 4.20 Numerical strain results for the laminate shown in Fig. 4.1, bending load case

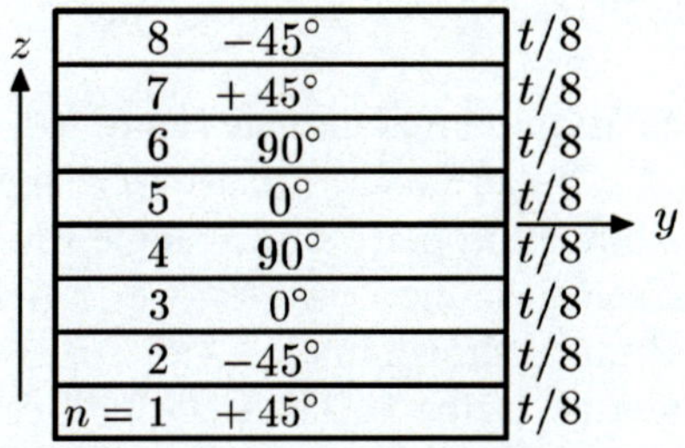

Fig. 4.21 Asymmetric laminate: $[\pm 45°/0°/90°/0°/90°/\pm 45°]$

Consider (a) a pure tensile load case with $N_x^{\text{n}} = 1000$ N/mm (all other forces and moments are equal to zero) and (b) a pure bending load case with $M_x^{\text{n}} = 1000$ Nmm/mm (all other moments and forces are equal to zero) to determine the stresses and strains in each lamina in the global x–y coordinate system as well as in the principal material coordinates (*1–2* coordinate systems) of each lamina.

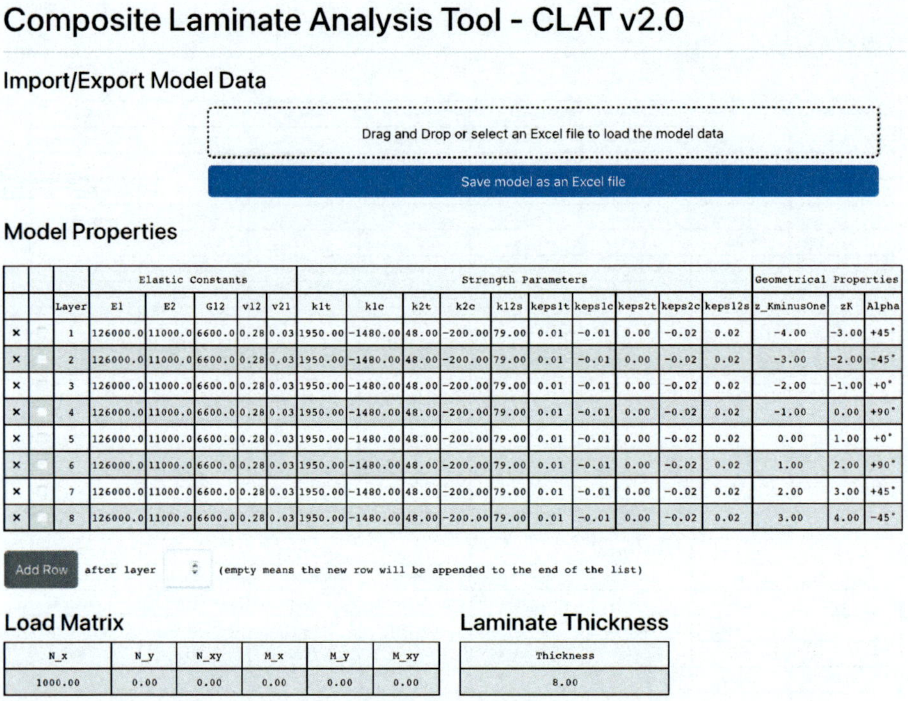

Composite Laminate Analysis Tool - CLAT v2.0

Import/Export Model Data

Drag and Drop or select an Excel file to load the model data

Save model as an Excel file

Model Properties

			Elastic Constants					Strength Parameters										Geometrical Properties		
		Layer	E1	E2	G12	v12	v21	k1t	k1c	k2t	k2c	k12s	keps1t	keps1c	keps2t	keps2c	keps12s	z_KminusOne	zK	Alpha
×		1	126000.0	11000.0	6600.0	0.28	0.03	1950.00	-1480.00	48.00	-200.00	79.00	0.01	-0.01	0.00	-0.02	0.02	-4.00	-3.00	+45°
×		2	126000.0	11000.0	6600.0	0.28	0.03	1950.00	-1480.00	48.00	-200.00	79.00	0.01	-0.01	0.00	-0.02	0.02	-3.00	-2.00	-45°
×		3	126000.0	11000.0	6600.0	0.28	0.03	1950.00	-1480.00	48.00	-200.00	79.00	0.01	-0.01	0.00	-0.02	0.02	-2.00	-1.00	+0°
×		4	126000.0	11000.0	6600.0	0.28	0.03	1950.00	-1480.00	48.00	-200.00	79.00	0.01	-0.01	0.00	-0.02	0.02	-1.00	0.00	+90°
×		5	126000.0	11000.0	6600.0	0.28	0.03	1950.00	-1480.00	48.00	-200.00	79.00	0.01	-0.01	0.00	-0.02	0.02	0.00	1.00	+0°
×		6	126000.0	11000.0	6600.0	0.28	0.03	1950.00	-1480.00	48.00	-200.00	79.00	0.01	-0.01	0.00	-0.02	0.02	1.00	2.00	+90°
×		7	126000.0	11000.0	6600.0	0.28	0.03	1950.00	-1480.00	48.00	-200.00	79.00	0.01	-0.01	0.00	-0.02	0.02	2.00	3.00	+45°
×		8	126000.0	11000.0	6600.0	0.28	0.03	1950.00	-1480.00	48.00	-200.00	79.00	0.01	-0.01	0.00	-0.02	0.02	3.00	4.00	-45°

Add Row after layer (empty means the new row will be appended to the end of the list)

Load Matrix

N_x	N_y	N_xy	M_x	M_y	M_xy
1000.00	0.00	0.00	0.00	0.00	0.00

Laminate Thickness

Thickness
8.00

Fig. 4.22 Model definition in CLAT for the laminate shown in Fig. 4.21, tensile load case

- Solution for the laminate loaded by $N_x^{\mathrm{n}} = 1000\,\mathrm{N/mm}$

The model definition of the asymmetric laminate [±45°/0°/90°/0°/90°/ ± 45°] in CLAT is shown in Fig. 4.22. It can be seen that every single layer is represented with its material and geometrical properties by a single row. The applied load, i.e., $N_x^{\mathrm{n}} = 1000\,\mathrm{N/mm}$, is represented in the `Load Matrix` field, and the thickness of the entire laminate ($t = 8$ mm) is shown in the `Laminate Thickness` field.

The graphical representation of the [±45°/0°/90°/0°/90°/ ± 45°] laminate in CLAT is given in Fig. 4.23. Layer numbers and angles of rotation are provided, and each angle is represented by a different color.

As the first result option, CLAT can display the stress distribution over the laminate thickness. For this option, the stresses can be displayed in the lamina-fixed *1–2* coordinate system (see Fig. 2.2) or in a global x–y coordinate system (see Fig. 1.2) of the entire laminate. The stress distributions over the laminate thickness shown in Figs. 4.24, 4.25, and 4.26 provide for this particular case, i.e., an asymmetric laminate and symmetric load case, asymmetric stress distributions in the local (*1–2*) as well as global (x–y) coordinate systems. Clearly, different distributions are obtained in the local compared to the global system. It can be seen in addition

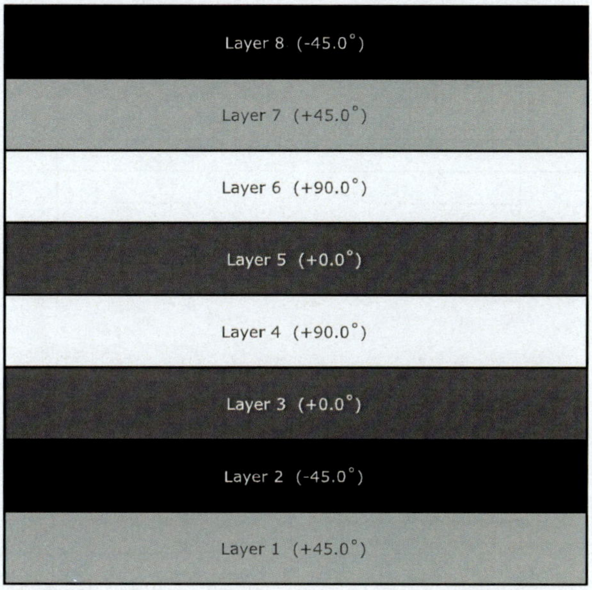

Fig. 4.23 Layer representation in CLAT for the laminate shown in Fig. 4.21

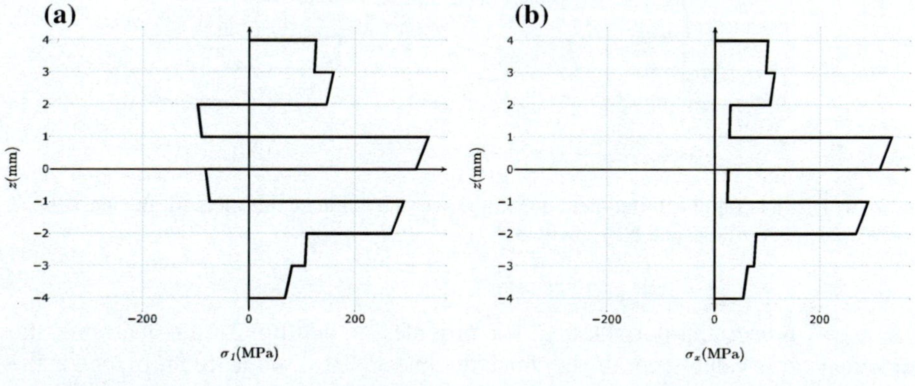

Fig. 4.24 Distribution of selected field quantities over the laminate thickness for the tensile load case: a normal stress $\sigma_I(z)$ and **b** normal stress $\sigma_x(z)$

that the uniaxial load case (i.e., $N_x^{\mathrm{n}} = 1000\,\mathrm{N/mm}$) results in a multiaxial stress state with other considerable normal and shear stress components; see Figs. 4.25 and 4.26. Finally, it can be stated that the considered case results in linear stress values within every single lamina.

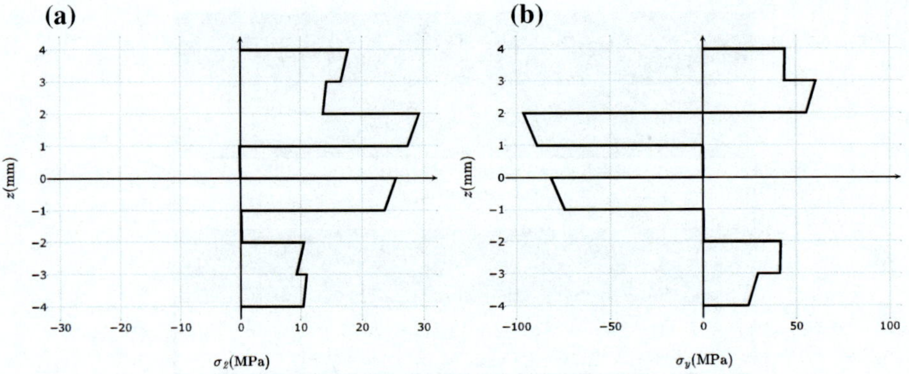

Fig. 4.25 Distribution of selected field quantities over the laminate thickness for the tensile load case: **a** normal stress $\sigma_2(z)$ and **b** normal stress $\sigma_y(z)$

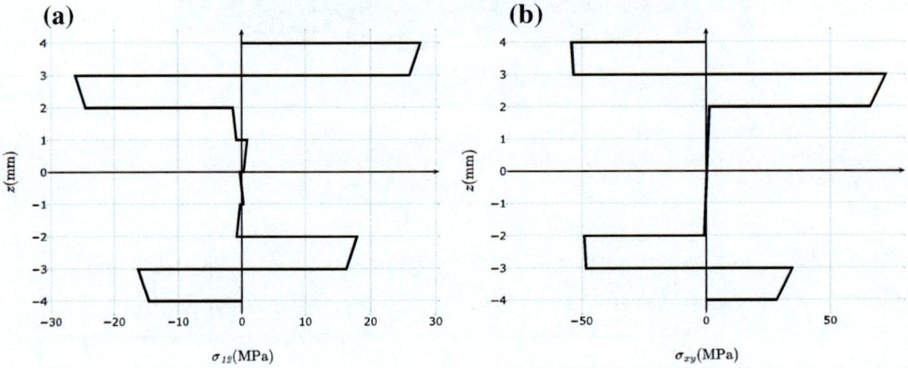

Fig. 4.26 Distribution of selected field quantities over the laminate thickness for the tensile load case: **a** shear stress $\sigma_{12}(z)$ and **b** shear stress $\sigma_{xy}(z)$

For post-processing purposes, CLAT provides, in addition to the diagrams, the numerical stress values over the laminate thickness. These values are provided at the bottom, in the middle, and at the top of the laminate; see Fig. 4.27.

As the second result option, CLAT can display the strain distribution over the laminate thickness. For this option, the strains can be displayed in the lamina-fixed *1–2* coordinate system (see Fig. 2.2) or in a global *x–y* coordinate system (see Fig. 1.2) of the entire laminate. The strain distributions over the laminate thickness shown in Figs. 4.28, 4.29, and 4.30 provide for this particular case, i.e., an asymmetric laminate and symmetric load case, asymmetric strain distributions in the local (*1–2*) as well as global (*x–y*) coordinate systems. However, the global coordinate representations provide a linear strain distribution across the entire laminate thickness.

Stress Table

Export

Layer no.	Position	sigma_1	sigma_2	sigma_12	sigma_x	sigma_y	sigma_xy
1	Bottom	67.561	10.423	-14.649	53.641	24.344	28.569
1	Center	74.089	10.692	-15.464	57.855	26.927	31.698
1	Top	80.617	10.962	-16.279	62.069	29.510	34.828
2	Bottom	105.675	9.346	16.279	73.790	41.231	-48.165
2	Center	107.102	9.944	17.095	75.618	41.429	-48.579
2	Top	108.530	10.542	17.910	77.446	41.626	-48.994
3	Bottom	268.947	0.199	-0.793	268.947	0.199	-0.793
3	Center	280.565	0.140	-0.520	280.565	0.140	-0.520
3	Top	292.184	0.081	-0.248	292.184	0.081	-0.248
4	Bottom	-74.071	23.695	0.248	23.695	-74.071	-0.248
4	Center	-77.734	24.622	-0.024	24.622	-77.734	0.024
4	Top	-81.397	25.548	-0.296	25.548	-81.397	0.296
5	Bottom	315.421	-0.037	0.296	315.421	-0.037	0.296
5	Center	327.039	-0.096	0.568	327.039	-0.096	0.568
5	Top	338.658	-0.155	0.840	338.658	-0.155	0.840
6	Bottom	-88.724	27.400	-0.840	27.400	-88.724	0.840
6	Center	-92.387	28.327	-1.112	28.327	-92.387	1.112
6	Top	-96.050	29.253	-1.384	29.253	-96.050	1.384
7	Bottom	145.897	13.653	-24.432	104.208	55.343	66.122
7	Center	152.425	13.922	-25.248	108.421	57.926	69.251
7	Top	158.953	14.191	-26.063	112.635	60.509	72.381
8	Bottom	122.802	16.522	26.063	95.725	43.599	-53.140
8	Center	124.229	17.120	26.878	97.553	43.796	-53.554
8	Top	125.656	17.718	27.694	99.381	43.993	-53.969

Fig. 4.27 Numerical stress results for the laminate shown in Fig. 4.1, tensile load case

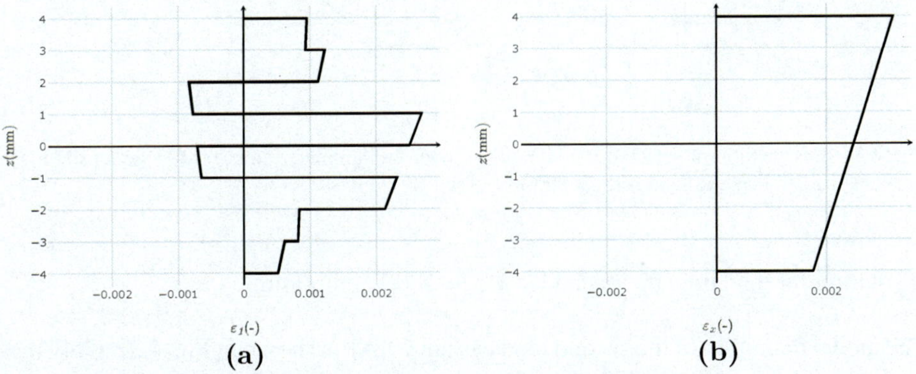

Fig. 4.28 Distribution of selected field quantities over the laminate thickness for the tensile load case: **a** normal strain $\epsilon_I(z)$ and **b** normal strain $\epsilon_x(z)$

As in the case of the stress distribution, numerical values for the strains can be obtained at the bottom, in the middle, and at the top of each lamina; see Fig. 4.31.

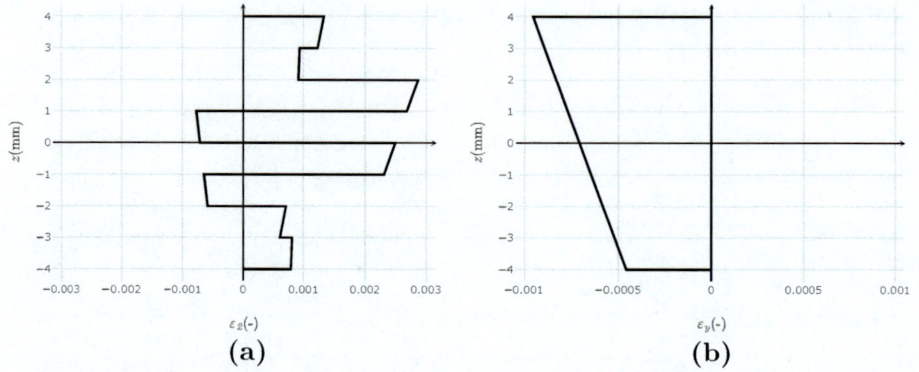

Fig. 4.29 Distribution of selected field quantities over the laminate thickness for the tensile load case: **a** normal strain $\epsilon_2(z)$ and **b** normal strain $\epsilon_y(z)$

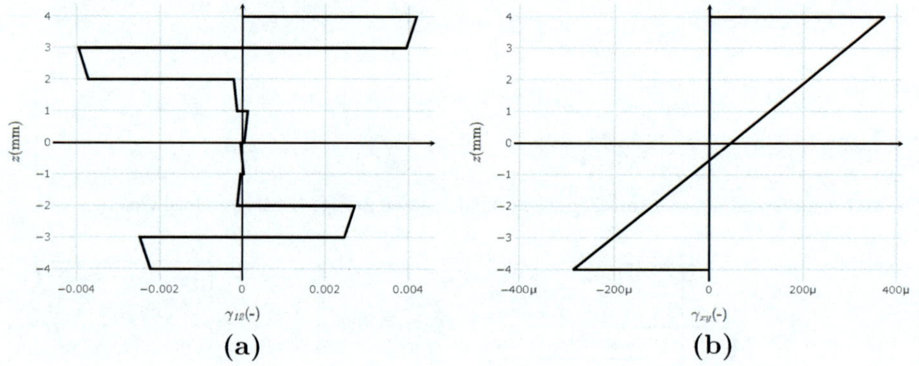

Fig. 4.30 Distribution of selected field quantities over the laminate thickness for the tensile load case: **a** shear strain $\gamma_{12}(z)$ and **b** shear strain $\gamma_{xy}(z)$

- Solution for the laminate loaded by $M_x^{\mathrm{n}} = 1000$ Nmm/mm

The model definition of the second load case in CLAT is shown in Fig. 4.32. Only the applied load (i.e., $M_x^{\mathrm{n}} = 1000$ Nmm/mm) must be modified in the `Load Matrix` field, and all other material and geometrical properties remain unchanged.

The stress distributions over the laminate thickness are shown in Figs. 4.33, 4.34, and 4.35 in the lamina-fixed *1–2* coordinate system (see Fig. 2.2) or in a global x–y coordinate system (see Fig. 1.2) of the entire laminate. It can be seen that for this particular case, i.e., an asymmetric laminate and bending load case, asymmetric stress distributions are obtained in both coordinate systems.

Strain Table

Export

Layer no.	Position	eps_1	eps_2	gamma_12	eps_x	eps_y	gamma_xy
1	Bottom	0.000512	0.000797	−0.002219	0.001765	−0.000455	−0.000285
1	Center	0.000564	0.000807	−0.002343	0.001857	−0.000486	−0.000244
1	Top	0.000615	0.000817	−0.002467	0.001949	−0.000517	−0.000203
2	Bottom	0.000817	0.000615	0.002467	0.001949	−0.000517	−0.000203
2	Center	0.000827	0.000666	0.002590	0.002042	−0.000548	−0.000161
2	Top	0.000837	0.000717	0.002714	0.002134	−0.000580	−0.000120
3	Bottom	0.002134	−0.000580	−0.000120	0.002134	−0.000580	−0.000120
3	Center	0.002226	−0.000611	−0.000079	0.002226	−0.000611	−0.000079
3	Top	0.002319	−0.000642	−0.000038	0.002319	−0.000642	−0.000038
4	Bottom	−0.000642	0.002319	0.000038	0.002319	−0.000642	−0.000038
4	Center	−0.000673	0.002411	−0.000004	0.002411	−0.000673	0.000004
4	Top	−0.000704	0.002503	−0.000045	0.002503	−0.000704	0.000045
5	Bottom	0.002503	−0.000704	0.000045	0.002503	−0.000704	0.000045
5	Center	0.002596	−0.000736	0.000086	0.002596	−0.000736	0.000086
5	Top	0.002688	−0.000767	0.000127	0.002688	−0.000767	0.000127
6	Bottom	−0.000767	0.002688	−0.000127	0.002688	−0.000767	0.000127
6	Center	−0.000798	0.002780	−0.000169	0.002780	−0.000798	0.000169
6	Top	−0.000829	0.002873	−0.000210	0.002873	−0.000829	0.000210
7	Bottom	0.001127	0.000917	−0.003702	0.002873	−0.000829	0.000210
7	Center	0.001178	0.000927	−0.003825	0.002965	−0.000860	0.000251
7	Top	0.001229	0.000937	−0.003949	0.003057	−0.000891	0.000292
8	Bottom	0.000937	0.001229	0.003949	0.003057	−0.000891	0.000292
8	Center	0.000947	0.001280	0.004072	0.003150	−0.000923	0.000333
8	Top	0.000957	0.001332	0.004196	0.003242	−0.000954	0.000375

Fig. 4.31 Numerical strain results for the laminate shown in Fig. 4.1, tensile load case

Composite Laminate Analysis Tool - CLAT v2.0

Import/Export Model Data

Drag and Drop or select an Excel file to load the model data

Save model as an Excel file

Model Properties

			Elastic Constants					Strength Parameters										Geometrical Properties		
		Layer	E1	E2	G12	v12	v21	k1t	k1c	k2t	k2c	k12s	keps1t	keps1c	keps2t	keps2c	keps12s	z_KminusOne	zK	Alpha
×		1	126000.0	11000.0	6600.0	0.28	0.03	1950.00	-1480.00	48.00	-200.00	79.00	0.01	-0.01	0.00	-0.02	0.02	-4.00	-3.00	+45°
×		2	126000.0	11000.0	6600.0	0.28	0.03	1950.00	-1480.00	48.00	-200.00	79.00	0.01	-0.01	0.00	-0.02	0.02	-3.00	-2.00	-45°
×		3	126000.0	11000.0	6600.0	0.28	0.03	1950.00	-1480.00	48.00	-200.00	79.00	0.01	-0.01	0.00	-0.02	0.02	-2.00	-1.00	+0°
×		4	126000.0	11000.0	6600.0	0.28	0.03	1950.00	-1480.00	48.00	-200.00	79.00	0.01	-0.01	0.00	-0.02	0.02	-1.00	0.00	+90°
×		5	126000.0	11000.0	6600.0	0.28	0.03	1950.00	-1480.00	48.00	-200.00	79.00	0.01	-0.01	0.00	-0.02	0.02	0.00	1.00	+0°
×		6	126000.0	11000.0	6600.0	0.28	0.03	1950.00	-1480.00	48.00	-200.00	79.00	0.01	-0.01	0.00	-0.02	0.02	1.00	2.00	+90°
×		7	126000.0	11000.0	6600.0	0.28	0.03	1950.00	-1480.00	48.00	-200.00	79.00	0.01	-0.01	0.00	-0.02	0.02	2.00	3.00	+45°
×		8	126000.0	11000.0	6600.0	0.28	0.03	1950.00	-1480.00	48.00	-200.00	79.00	0.01	-0.01	0.00	-0.02	0.02	3.00	4.00	-45°

Add Row after layer (empty means the new row will be appended to the end of the list)

Load Matrix

N_x	N_y	N_xy	M_x	M_y	M_xy
0.00	0.00	0.00	1000.00	0.00	0.00

Laminate Thickness

Thickness
8.00

Fig. 4.32 Model definition in CLAT for the laminate shown in Fig. 4.21, bending load case

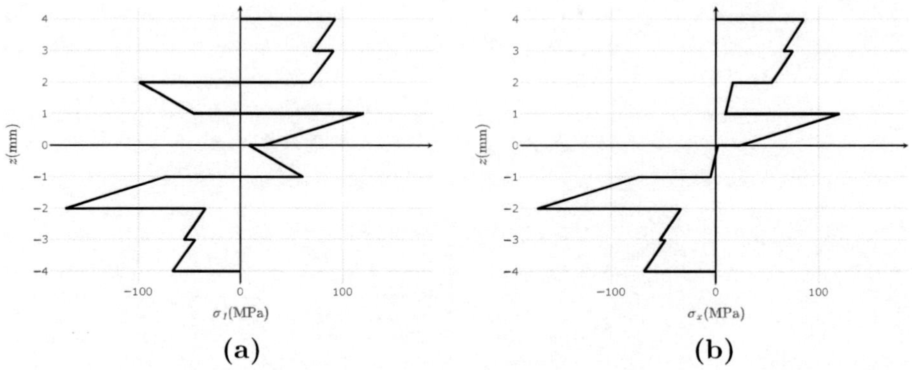

Fig. 4.33 Distribution of selected field quantities over the laminate thickness for the bending load case: **a** normal stress $\sigma_1(z)$ and **(b)** normal stress $\sigma_x(z)$

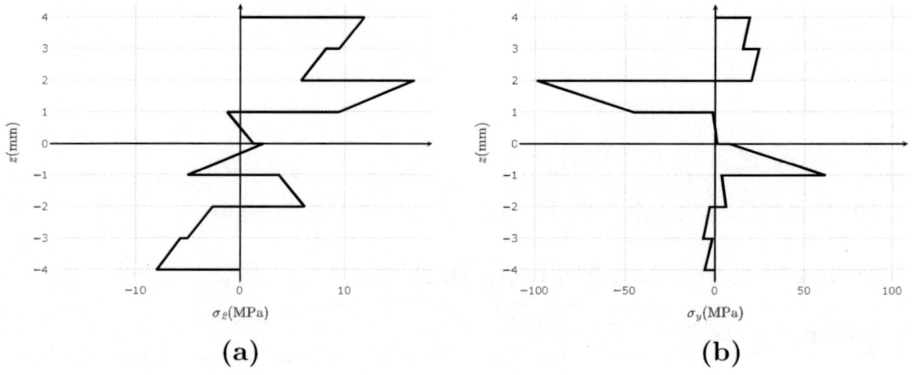

Fig. 4.34 Distribution of selected field quantities over the laminate thickness for the bending load case: **a** normal stress $\sigma_2(z)$ and **b** normal stress $\sigma_y(z)$

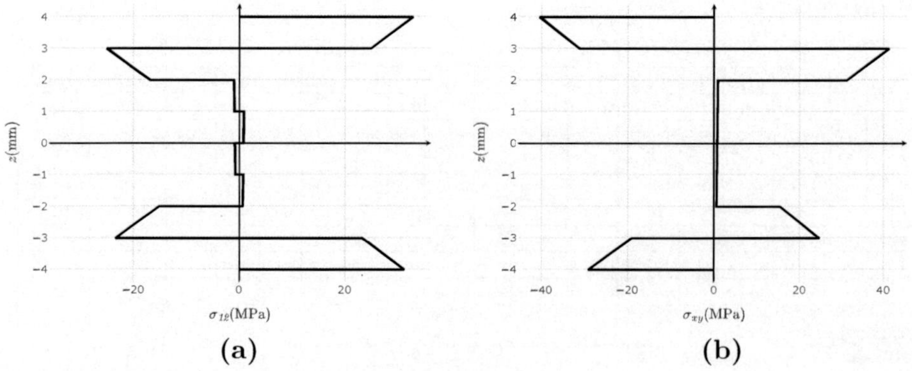

Fig. 4.35 Distribution of selected field quantities over the laminate thickness for the bending load case: **a** shear stress $\sigma_{12}(z)$ and **b** shear stress $\sigma_{xy}(z)$

Stress Table

Export

Layer no.	Position	sigma_1	sigma_2	sigma_12	sigma_x	sigma_y	sigma_xy
1	Bottom	-66.107	-8.035	31.367	-68.439	-5.704	-29.036
1	Center	-54.891	-6.873	27.345	-58.228	-3.537	-24.009
1	Top	-43.675	-5.711	23.323	-48.017	-1.370	-18.982
2	Bottom	-54.631	-5.005	-23.323	-53.142	-6.495	24.813
2	Center	-44.115	-3.798	-19.301	-43.258	-4.655	20.158
2	Top	-33.598	-2.591	-15.279	-33.374	-2.815	15.504
3	Bottom	-170.614	6.244	0.659	-170.614	6.244	0.659
3	Center	-122.055	4.998	0.696	-122.055	4.998	0.696
3	Top	-73.496	3.752	0.734	-73.496	3.752	0.734
4	Bottom	62.121	-4.992	-0.734	-4.992	62.121	0.734
4	Center	35.295	-1.377	-0.771	-1.377	35.295	0.771
4	Top	8.469	2.238	-0.808	2.238	8.469	0.808
5	Bottom	23.622	1.261	0.808	23.622	1.261	0.808
5	Center	72.181	0.015	0.846	72.181	0.015	0.846
5	Top	120.740	-1.231	0.883	120.740	-1.231	0.883
6	Bottom	-45.183	9.467	-0.883	9.467	-45.183	0.883
6	Center	-72.009	13.082	-0.920	13.082	-72.009	0.920
6	Top	-98.835	16.697	-0.958	16.697	-98.835	0.958
7	Bottom	68.487	5.909	-16.896	54.094	20.301	31.289
7	Center	79.703	7.071	-20.918	64.305	22.468	36.316
7	Top	90.919	8.233	-24.940	74.516	24.636	41.343
8	Bottom	71.570	9.480	24.940	65.465	15.584	-31.045
8	Center	82.086	10.687	28.962	75.349	17.424	-35.699
8	Top	92.603	11.894	32.984	85.233	19.264	-40.354

Fig. 4.36 Numerical stress results for the laminate shown in Fig. 4.21, bending load case

For post-processing purposes, the numerical stress values over the laminate thickness, i.e., at the bottom, in the middle, and at the top of the laminate, are provided in Fig. 4.36.

The strain distributions over the laminate thickness are shown in Figs. 4.37, 4.38, and 4.39 **in the lamina-fixed** *1–2* coordinate system (see Fig. 2.2) or in a global x–y coordinate system (see Fig. 1.2) of the entire laminate. It can be seen that for this particular case, i.e., an asymmetric laminate and bending load case, complex asymmetric strain distributions are obtained in the local (*1–2*) system, whereas perfect straight lines through the coordinate origin are seen in the global (x–y) coordinate system.

For post-processing purposes, the numerical strain values over the laminate thickness, i.e., at the bottom, in the middle, and at the top of the laminate, are provided in Fig. 4.40.

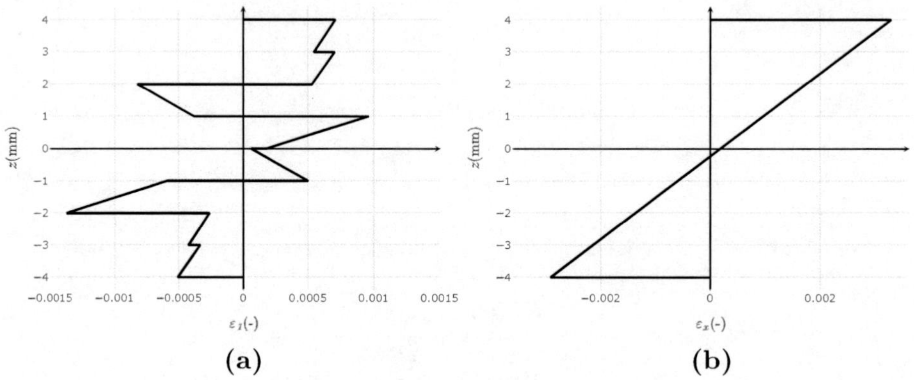

Fig. 4.37 Distribution of selected field quantities over the laminate thickness for the bending load case: **a** normal strain $\epsilon_1(z)$ and **b** normal strain $\epsilon_x(z)$

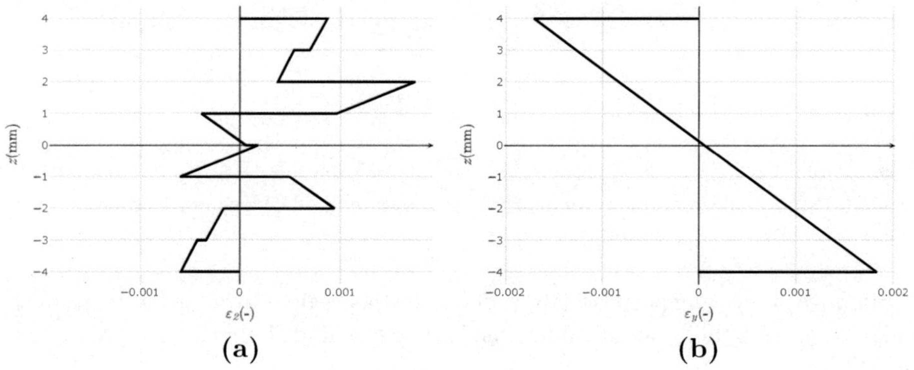

Fig. 4.38 Distribution of selected field quantities over the laminate thickness for the bending load case: **a** normal strain $\epsilon_2(z)$ and **b** normal strain $\epsilon_y(z)$

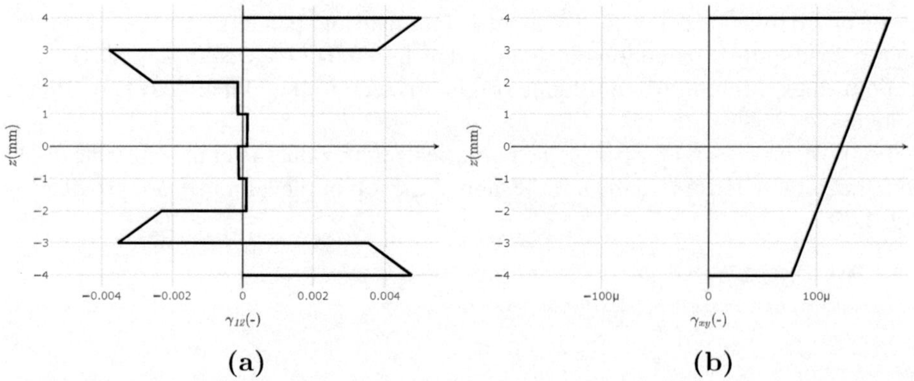

Fig. 4.39 Distribution of selected field quantities over the laminate thickness for the bending load case: **a** shear strain $\gamma_{12}(z)$ and **b** shear strain $\gamma_{xy}(z)$

Strain Table

Export

Layer no.	Position	eps_1	eps_2	gamma_12	eps_x	eps_y	gamma_xy
1	Bottom	−0.000506	−0.000584	0.004753	−0.002921	0.001831	0.000077
1	Center	−0.000420	−0.000503	0.004143	−0.002533	0.001610	0.000083
1	Top	−0.000334	−0.000422	0.003534	−0.002145	0.001389	0.000089
2	Bottom	−0.000422	−0.000334	−0.003534	−0.002145	0.001389	0.000089
2	Center	−0.000341	−0.000247	−0.002924	−0.001757	0.001168	0.000094
2	Top	−0.000261	−0.000161	−0.002315	−0.001368	0.000947	0.000100
3	Bottom	−0.001368	0.000947	0.000100	−0.001368	0.000947	0.000100
3	Center	−0.000980	0.000726	0.000106	−0.000980	0.000726	0.000106
3	Top	−0.000592	0.000504	0.000111	−0.000592	0.000504	0.000111
4	Bottom	0.000504	−0.000592	−0.000111	−0.000592	0.000504	0.000111
4	Center	0.000283	−0.000204	−0.000117	−0.000204	0.000283	0.000117
4	Top	0.000062	0.000185	−0.000122	0.000185	0.000062	0.000122
5	Bottom	0.000185	0.000062	0.000122	0.000185	0.000062	0.000122
5	Center	0.000573	−0.000159	0.000128	0.000573	−0.000159	0.000128
5	Top	0.000961	−0.000380	0.000134	0.000961	−0.000380	0.000134
6	Bottom	−0.000380	0.000961	−0.000134	0.000961	−0.000380	0.000134
6	Center	−0.000601	0.001349	−0.000139	0.001349	−0.000601	0.000139
6	Top	−0.000823	0.001738	−0.000145	0.001738	−0.000823	0.000145
7	Bottom	0.000530	0.000385	−0.002560	0.001738	−0.000823	0.000145
7	Center	0.000616	0.000466	−0.003169	0.002126	−0.001044	0.000151
7	Top	0.000703	0.000546	−0.003779	0.002514	−0.001265	0.000156
8	Bottom	0.000546	0.000703	0.003779	0.002514	−0.001265	0.000156
8	Center	0.000627	0.000789	0.004388	0.002902	−0.001486	0.000162
8	Top	0.000708	0.000876	0.004998	0.003290	−0.001707	0.000168

Fig. 4.40 Numerical strain results for the laminate shown in Fig. 4.21, bending load case

4.4 Problem 3: Failure Criteria

Given is an asymmetric 4-layer laminate of thickness $t = 8$ mm as shown in Fig. 4.41. Each lamina has the same thickness ($t/4$) and identical material properties, which can be taken from Table 4.1. However, the orientation of each lamina changes from layer to layer as indicated in Fig. 4.41.

Consider (a) a biaxial tensile load case with $N_x^{\text{n}} = 1000$ N/mm and $N_y^{\text{n}} = 1000$ N/mm (all other forces and moments are equal to zero) and (b) a biaxial bending load case with $M_x^{\text{n}} = 1000$ Nmm/mm and $M_y^{\text{n}} = 500$ Nmm/mm (all other moments and forces are equal to zero) to determine the stresses and strains in each lamina in the global x–y coordinate system as well as in the principal material coordinates (*1–2* coordinate systems) of each lamina. Use these stress and strain values to perform a failure analysis based on the maximum stress, maximum strain, Tsai-Hill, and Tsai-Wu failure criteria.

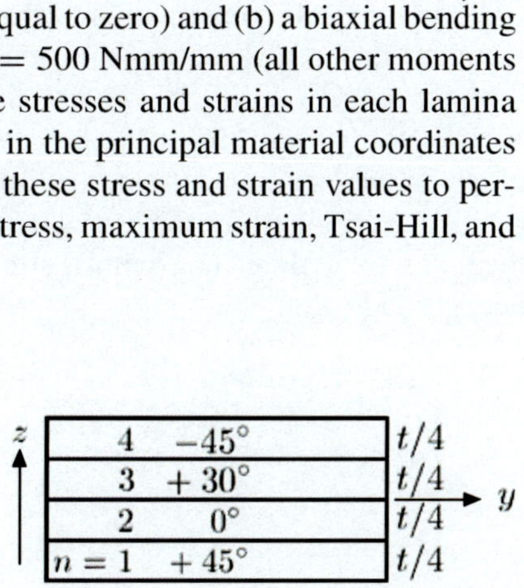

Fig. 4.41 Asymmetric laminate: [+45°/0°/+30°/−45°]

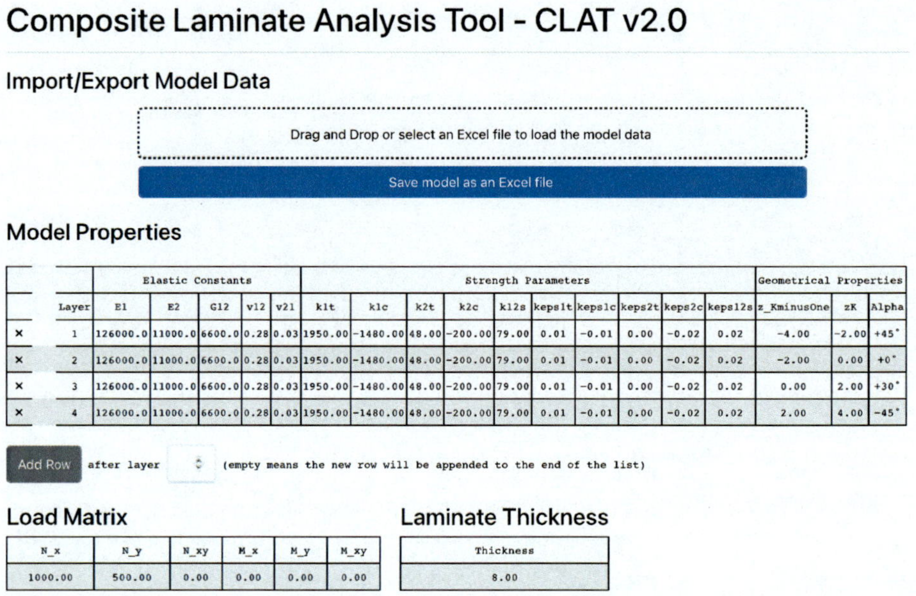

Fig. 4.42 **Model definition in CLAT for the laminate shown in Fig. 4.41**, tensile load case

- Solution for the laminate loaded by $N_x^{\text{n}} = 1000\,\text{N/mm}$ and $N_y^{\text{n}} = 500\,\text{N/mm}$

The model definition of the asymmetric laminate $[+45°/\,0°/+30°/-45°]$ in CLAT is shown in Fig. 4.42. It can be seen that every single layer is represented with its material and geometrical properties by a single row. The applied loads, i.e., $N_x^{\text{n}} = 1000$ N/mm and $N_y^{\text{n}} = 1000\,\text{N/mm}$, are represented in the `Load Matrix` field, and the thickness of the entire laminate ($t = 8$ mm) is shown in the `Laminate Thickness` field. It is important to note that the solution procedure requires setting the ☑`Failure Analysis` option.

The graphical representation of the $[+45°/\,0°/+30°/-45°]$ laminate in CLAT is given in Fig. 4.43. Layer numbers and angles of rotation are provided and each angle is represented by a different color.

The strength ratio R at the bottom, in the middle, and at the top of each layer based on the maximum stress, maximum strain, Tsai-Hill, and Tsai-Wu criteria can be displayed by using the `Failure Analysis` option; see Fig. 4.44. Critical locations for failure, i.e., with a strength ratio $R \leq 1$, are highlighted with a red background color.

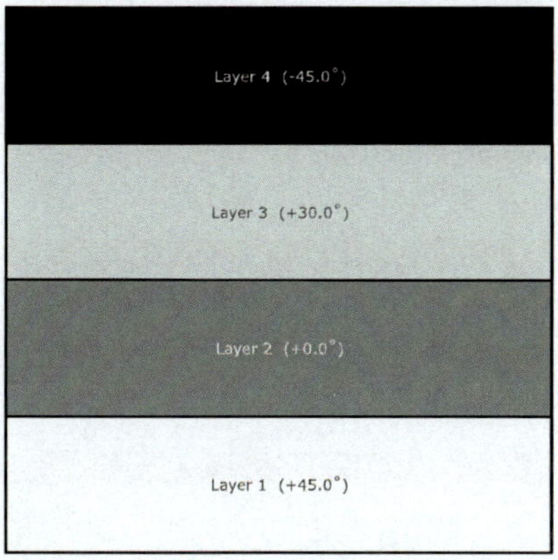

Fig. 4.43 Layer representation in CLAT for the laminate shown in Fig. 4.41

• Solution for the laminate loaded by M_x^{n}=1000 Nmm/mm and $M_y^{\text{n}} = 500$ Nmm/mm

The model definition of the second load case in CLAT is shown in Fig. 4.45. Only the applied loads (i.e., $M_x^{\text{n}} = 1000$ Nmm/mm and $M_y^{\text{n}} = 500$ Nmm/mm) must be modified in the `Load Matrix` field, and all other material and geometrical properties remain unchanged.

The strength ratios R for the second load case are shown in Fig. 4.46. It can be seen that no ratio is in the critical range, and thus, failure should not occur.

4.5 Problem 4: Ply-By-Ply Failure Loads

Given is a symmetric 3-layer laminate of thickness $t = 9$ mm as shown in Fig. 4.47. Each lamina has the same thickness ($t/3$) and identical material properties, which can be taken from Table 4.1. However, the orientation of each lamina changes from layer to layer as indicated in Fig. 4.47.

Consider a pure tensile load case with $N_x^{\text{n}} > 0$ (all other forces and moments are equal to zero) to investigate the ply-by-ply failure of this laminate based on the Tsai-Wu criterion.

• Solution

The model definition of the symmetric laminate $[0°/\overline{90°}]_{\text{s}}$ in CLAT is shown in Fig. 4.48. It can be seen that every single layer is represented with its material and geometrical properties by a single row. A unit load, i.e., $N_x^{\text{n}} = 1$ N/mm, is represented in the `Load Matrix` field, and the thickness of the entire laminate ($t = 9$ mm) is shown in the `Laminate Thickness` field. It is important to note that the solution procedure requires setting the ☑`Ply-By-Ply Analysis` option.

Export

			Strength Ration R			
Layer no.	Position	Direction	Stress Max	Strain Max	Tsai-Hill	Tsai-Wu
1	Bottom	1	100.093	2605.328	0.000	0.000
1	Bottom	2	0.689	0.693	0.681	0.680
1	Bottom	12	4.569	7.634	0.000	0.000
1	Middle	1	35.646	47.161	0.000	0.000
1	Middle	2	0.774	0.791	0.767	0.773
1	Middle	12	5.474	9.146	0.000	0.000
1	Top	1	21.684	23.400	0.000	0.000
1	Top	2	0.884	0.920	0.876	0.896
1	Top	12	6.826	11.406	0.000	0.000
2	Bottom	1	8.194	7.724	0.000	0.000
2	Bottom	2	1.072	1.232	0.999	1.037
2	Bottom	12	2.886	4.823	0.000	0.000
2	Middle	1	8.352	7.830	0.000	0.000
2	Middle	2	1.212	1.414	1.148	1.215
2	Middle	12	3.897	6.511	0.000	0.000
2	Top	1	8.517	7.938	0.000	0.000
2	Top	2	1.392	1.661	1.342	1.456
2	Top	12	5.997	10.020	0.000	0.000
3	Bottom	1	13.045	12.592	0.000	0.000
3	Bottom	2	1.212	1.334	1.208	1.286
3	Bottom	12	52.073	87.007	0.000	0.000
3	Middle	1	10.307	9.655	0.000	0.000
3	Middle	2	1.518	1.777	1.505	1.658
3	Middle	12	177.858	297.180	0.000	0.000
3	Top	1	8.519	7.829	0.000	0.000
3	Top	2	2.032	2.661	1.982	2.288
3	Top	12	125.643	209.935	0.000	0.000
4	Bottom	1	9.215	8.502	0.000	0.000
4	Bottom	2	1.941	2.452	1.904	2.170
4	Bottom	12	586.285	979.615	0.000	0.000
4	Middle	1	17.137	16.338	0.000	0.000
4	Middle	2	1.874	2.101	1.862	2.000
4	Middle	12	28.993	48.445	0.000	0.000
4	Top	1	122.180	208.456	0.000	0.000
4	Top	2	1.812	1.838	1.798	1.807
4	Top	12	14.147	23.638	0.000	0.000

Fig. 4.44 Numerical R values for the laminate shown in Fig. 4.41, tensile load case

The graphical representation of the $[0°/\overline{90°}]_s$ laminate in CLAT is given in Fig. 4.49. Layer numbers and angles of rotation are provided, and each angle is represented by a different color.

Figures 4.50, 4.51 and 4.52 display the stress and strain distributions in the x-direction for different steps of the ply-by-ply failure analysis. Step 0 represents the original configuration at which all layers are in good shape and take part of the load; see Fig. 4.50.

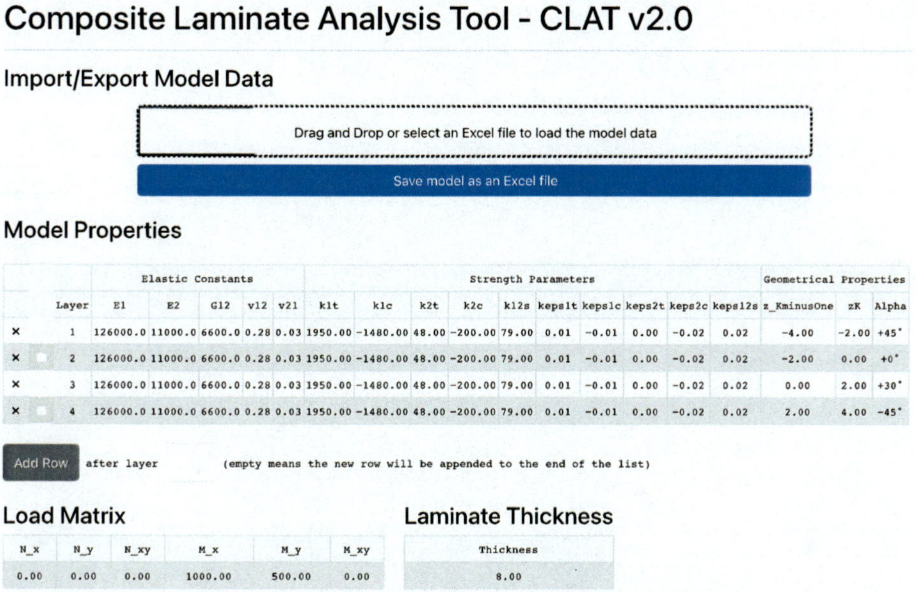

	Elastic Constants						Strength Parameters										Geometrical Properties		
	Layer	E1	E2	G12	v12	v21	k1t	k1c	k2t	k2c	k12s	keps1t	keps1c	keps2t	keps2c	keps12s	z_KminusOne	zK	Alpha
×	1	126000.0	11000.0	6600.0	0.28	0.03	1950.00	-1480.00	48.00	-200.00	79.00	0.01	-0.01	0.00	-0.02	0.02	-4.00	-2.00	+45°
×	2	126000.0	11000.0	6600.0	0.28	0.03	1950.00	-1480.00	48.00	-200.00	79.00	0.01	-0.01	0.00	-0.02	0.02	-2.00	0.00	+0°
×	3	126000.0	11000.0	6600.0	0.28	0.03	1950.00	-1480.00	48.00	-200.00	79.00	0.01	-0.01	0.00	-0.02	0.02	0.00	2.00	+30°
×	4	126000.0	11000.0	6600.0	0.28	0.03	1950.00	-1480.00	48.00	-200.00	79.00	0.01	-0.01	0.00	-0.02	0.02	2.00	4.00	-45°

N_x	N_y	N_xy	M_x	M_y	M_xy
0.00	0.00	0.00	1000.00	500.00	0.00

Thickness
8.00

Fig. 4.45 Model definition in CLAT for the laminate shown in Fig. 4.41, bending load case

Step 1 represents the situation with one layer, i.e., layer 2 already failed. Thus, the stress and strain distributions are zero in the range of layer 2, i.e., $-2 \leq z \leq +2$; see Fig. 4.51.

The final stage with all layers failed is represented by step 2; see Fig. 4.52. Now the stress and strain are everywhere across the thickness zero.

The macroscopic stress-strain diagram under consideration of the layer-wise failure of single laminae is presented in Fig. 4.53.

4.6 Problem 5: Pole Diagrams of Elastic Properties for Unidirectional Laminae

Given is a unidirectional lamina, i.e., a AS4 carbon/3501-6 epoxy prepreg, of 1 mm thickness; see Fig. 4.54.

Calculate the apparent elastic properties E_x, E_y, G_{xy}, ν_{xy}, and ν_{yx} as a function of the angle of rotation. Represent the elastic properties in the form of pole diagrams ($0° \leq \alpha \leq 360°$). In addition, provide a classical representation in a Cartesian coordinate system ($0° \leq \alpha \leq 90°$). The material properties of the lamina can be taken from Table 4.1.

Export

			Strength Ration R			
Layer no.	Position	Direction	Stress Max	Strain Max	Tsai-Hill	Tsai-Wu
1	Bottom	1	15.769	17.404	0.000	0.000
1	Bottom	2	6.546	7.785	5.270	6.652
1	Bottom	12	10.018	16.739	0.000	0.000
1	Middle	1	23.968	27.122	0.000	0.000
1	Middle	2	8.032	9.405	6.956	8.301
1	Middle	12	15.568	26.013	0.000	0.000
1	Top	1	49.928	61.416	0.000	0.000
1	Top	2	10.391	11.877	9.890	10.813
1	Top	12	34.907	58.326	0.000	0.000
2	Bottom	1	10.337	10.594	0.000	0.000
2	Bottom	2	16.767	26.103	6.010	9.824
2	Bottom	12	8.019	13.399	0.000	0.000
2	Middle	1	20.465	21.212	0.000	0.000
2	Middle	2	22.793	31.399	7.987	11.936
2	Middle	12	9.264	15.479	0.000	0.000
2	Top	1	1013.161	11144.061	0.000	0.000
2	Top	2	35.581	39.389	10.482	13.903
2	Top	12	10.966	18.324	0.000	0.000
3	Bottom	1	47.256	39.813	0.000	0.000
3	Bottom	2	23.878	23.443	20.544	19.078
3	Bottom	12	114.058	190.578	0.000	0.000
3	Middle	1	22.108	19.479	0.000	0.000
3	Middle	2	54.344	37.695	17.160	18.249
3	Middle	12	32.864	54.911	0.000	0.000
3	Top	1	14.429	12.894	0.000	0.000
3	Top	2	47.279	96.141	9.969	10.066
3	Top	12	14.363	23.998	0.000	0.000
4	Bottom	1	145.743	160.561	0.000	0.000
4	Bottom	2	5.414	5.616	4.610	4.387
4	Bottom	12	8.796	14.696	0.000	0.000
4	Middle	1	27.739	26.115	0.000	0.000
4	Middle	2	3.716	4.282	3.231	3.234
4	Middle	12	6.699	11.193	0.000	0.000
4	Top	1	15.328	14.214	0.000	0.000
4	Top	2	2.828	3.460	2.478	2.547
4	Top	12	5.409	9.038	0.000	0.000

Fig. 4.46 Numerical R values for the laminate shown in Fig. 4.41, bending load case

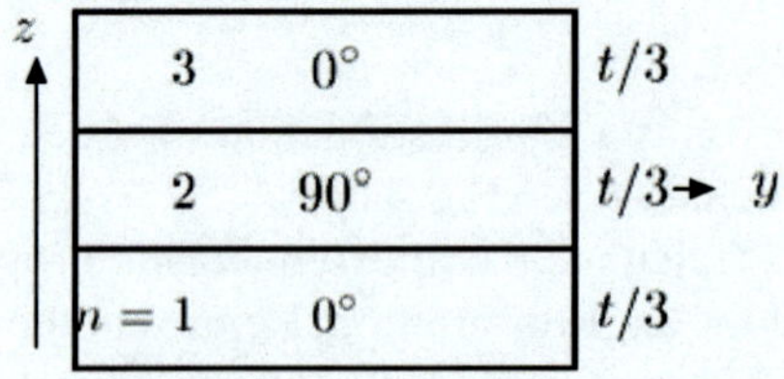

Fig. 4.47 Symmetric laminate: $[0°/\overline{90°}]_s$

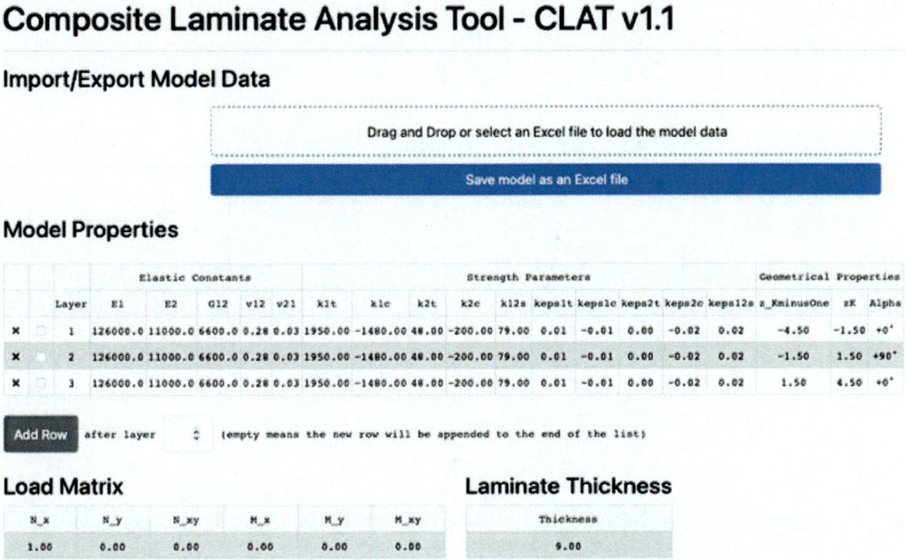

Fig. 4.48 Model definition in CLAT for the laminate shown in Fig. 4.47, ply-by-ply failure

Fig. 4.49 Layer representation in CLAT for the laminate shown in Fig. 4.47

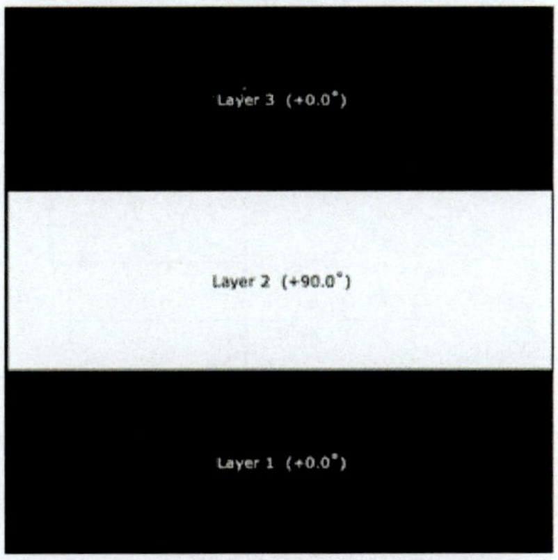

• Solution

The model definition of the unidirectional lamina in CLAT is shown in Fig. 4.55. It can be seen that a single layer is represented with its material and geometrical properties by a single row. The applied load,[2] i.e., $N_x^n = 1000\,\text{N/mm}$, is represented in the `Load Matrix` field, and the thickness of the entire laminate ($t = 1$ mm) is shown in the `Laminate Thickness` field.

---

[2] Any load value can be provided here since the calculation is *only* about elastic constants.

Step 0

This is the original configuration of the problem.

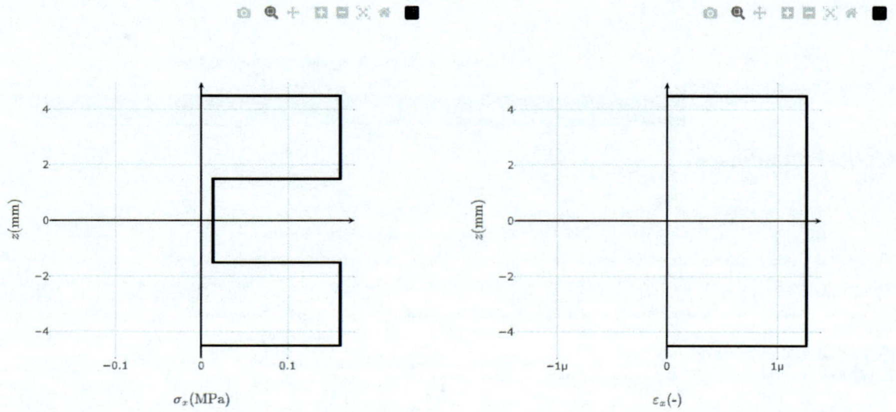

Fig. 4.50 Stress and strain distributions in the x-direction for the laminate shown in Fig. 4.47, ply-by-ply failure (step 0)

Step 1

With the strength ratio of 3469.718, failure occurred for layer number(s) [2].

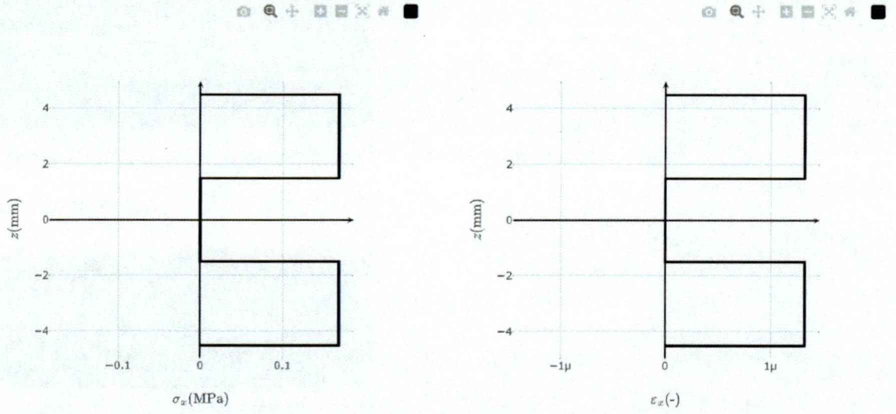

At the end of this step, the following layers have already been failed: [2]

Fig. 4.51 Stress and strain distributions in the x-direction for the laminate shown in Fig. 4.47, ply-by-ply failure (step 1)

Step 2

With the strength ratio of 11700.000, failure occurred for layer number(s) [1, 3].

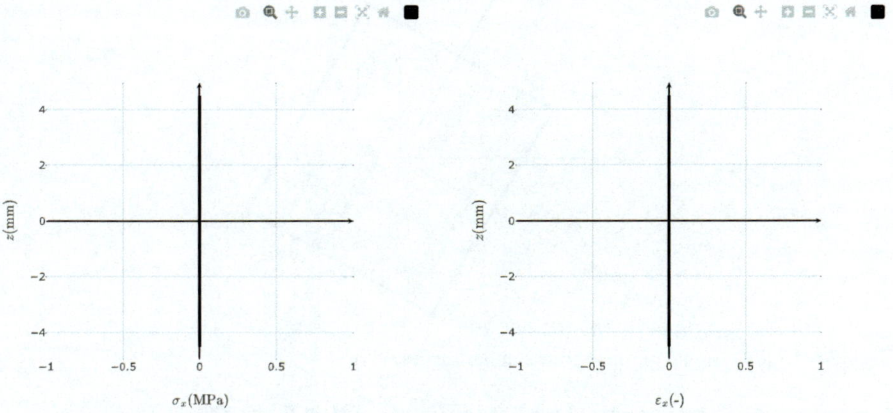

At the end of this step, the following layers have already been failed: [1, 2, 3]

Fig. 4.52 Stress and strain distributions in the x-direction for the laminate shown in Fig. 4.47, ply-by-ply failure (step 2)

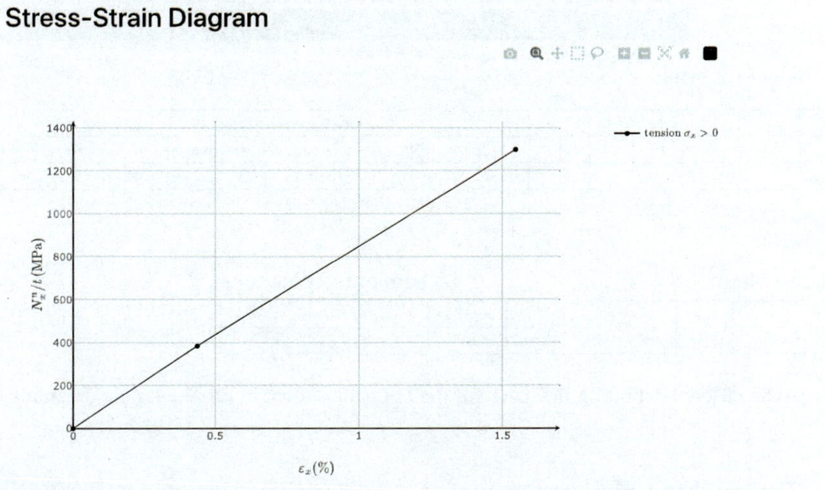

Fig. 4.53 Macroscopic stress-strain diagram to illustrate the ply-by-ply failure of the laminate shown in Fig. 4.47

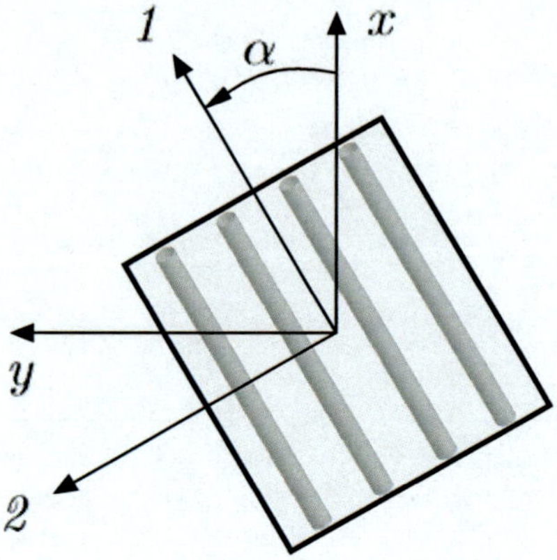

Fig. 4.54 Unidirectional lamina rotated under the angle α

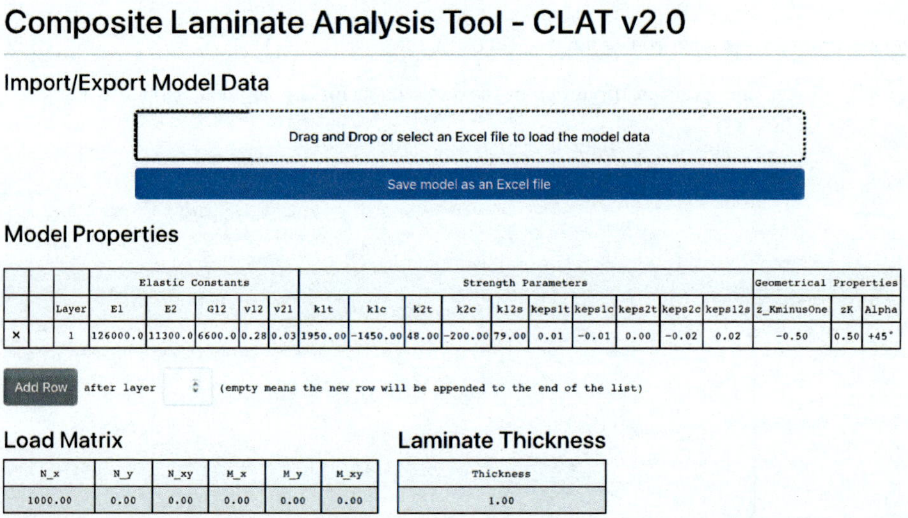

Composite Laminate Analysis Tool - CLAT v2.0

Import/Export Model Data

Drag and Drop or select an Excel file to load the model data

Save model as an Excel file

Model Properties

			Elastic Constants					Strength Parameters										Geometrical Properties		
		Layer	E1	E2	G12	v12	v21	k1t	k1c	k2t	k2c	k12s	keps1t	keps1c	keps2t	keps2c	keps12s	z_KminusOne	zK	Alpha
×		1	126000.0	11300.0	6600.0	0.28	0.03	1950.00	-1450.00	48.00	-200.00	79.00	0.01	-0.01	0.00	-0.02	0.02	-0.50	0.50	+45°

Add Row after layer (empty means the new row will be appended to the end of the list)

Load Matrix

N_x	N_y	N_xy	M_x	M_y	M_xy
1000.00	0.00	0.00	0.00	0.00	0.00

Laminate Thickness

Thickness
1.00

Fig. 4.55 Model definition in CLAT for the laminate shown in Fig. 4.54, pole diagrams

The graphical representation of the unidirectional lamina in CLAT is given in Fig. 4.56. Layer number and angle of rotation are provided.

The graphical representations of the apparent elastic properties for the AS4 carbon/3501-6 epoxy prepreg are shown in Figs. 4.57, 4.58, 4.59, and 4.60 as polar (pole) diagrams and in Fig. 4.61 in a classical Cartesian representation.

Fig. 4.56 Layer representation in CLAT for the unidirectional lamina shown in Fig. 4.54

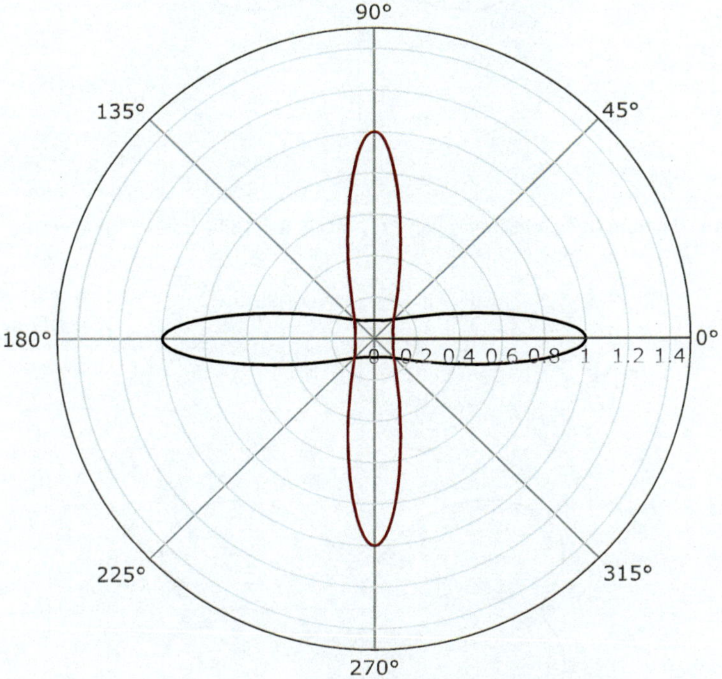

Fig. 4.57 Pole diagram of the elastic properties E_x/E_1 (black) and E_y/E_1 (red) for AS4 carbon/3501-6 epoxy

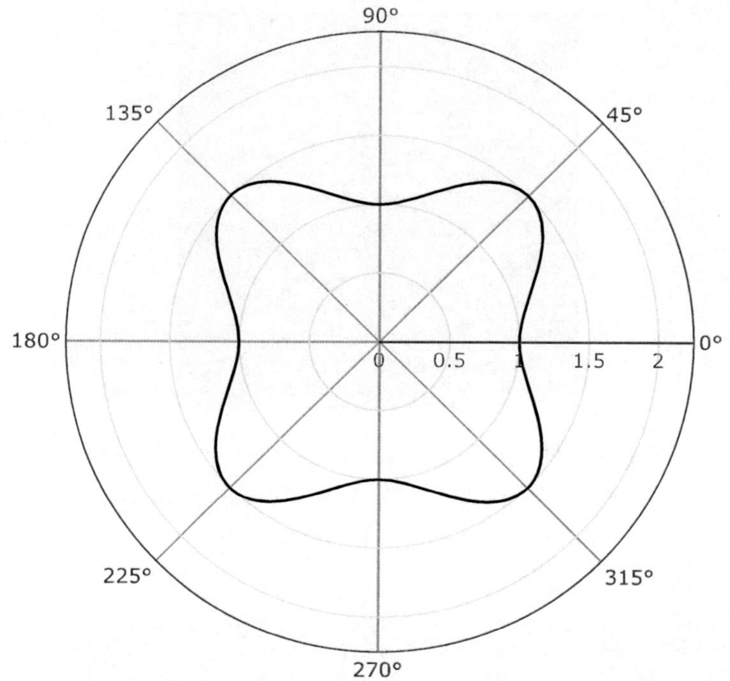

Fig. 4.58 Pole diagram of the elastic property G_{xy}/G_{12} for AS4 carbon/3501-6 epoxy

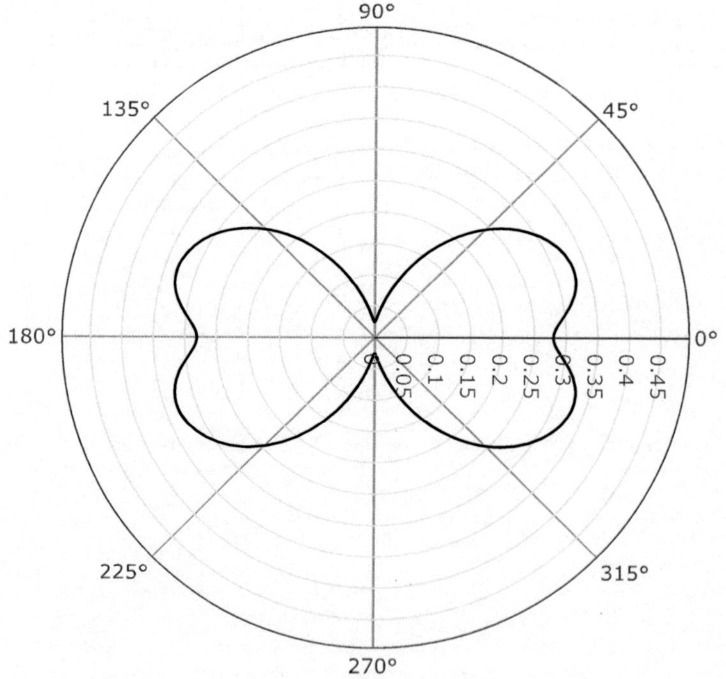

Fig. 4.59 Pole diagram of the elastic property ν_{xy} for AS4 carbon/3501-6 epoxy

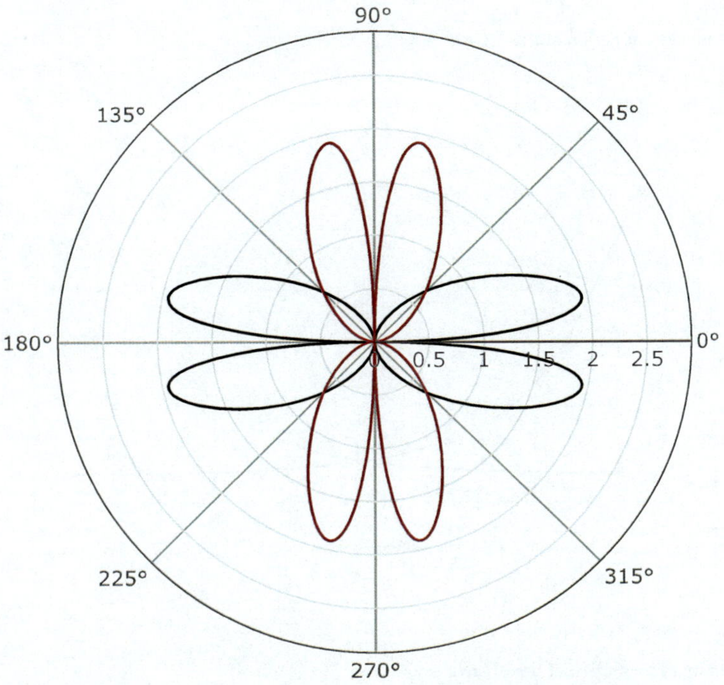

Fig. 4.60 Pole diagram of the elastic properties $-\eta_x$ (black) and $-\eta_y$ (red) for AS4 carbon/3501-6 epoxy

4.7 Problem 6: Failure Envelopes for Unidirectional Laminae

Given is a unidirectional lamina, i.e., a AS4 carbon/3501-6 epoxy prepreg, of 1 mm thickness, which is loaded by a normal stress σ_x in the x-direction; see Fig. 4.62.

Calculate and sketch the failure envelope, i.e., the failure stress in the x-direction as a function of the angle of rotation ($0° \leq \alpha \leq 90°$), based on the maximum stress, maximum strain, Tsai-Hill, and Tsai-Wu failure criteria. The material properties of the lamina can be taken from Table 4.1.

• Solution

The model definition of the unidirectional lamina in CLAT is shown in Fig. 4.63. It can be seen that a single layer is represented with its material and geometrical properties by a single row. The applied load,[3] i.e., $N_x^{\text{n}} = 1000\,\text{N/mm}$, is represented in the `Load Matrix` field, and the thickness of the entire laminate ($t = 1$ mm) is shown in the `Laminate Thickness` field.

[3] Any load value can be provided here since the calculation is *only* about elastic constants.

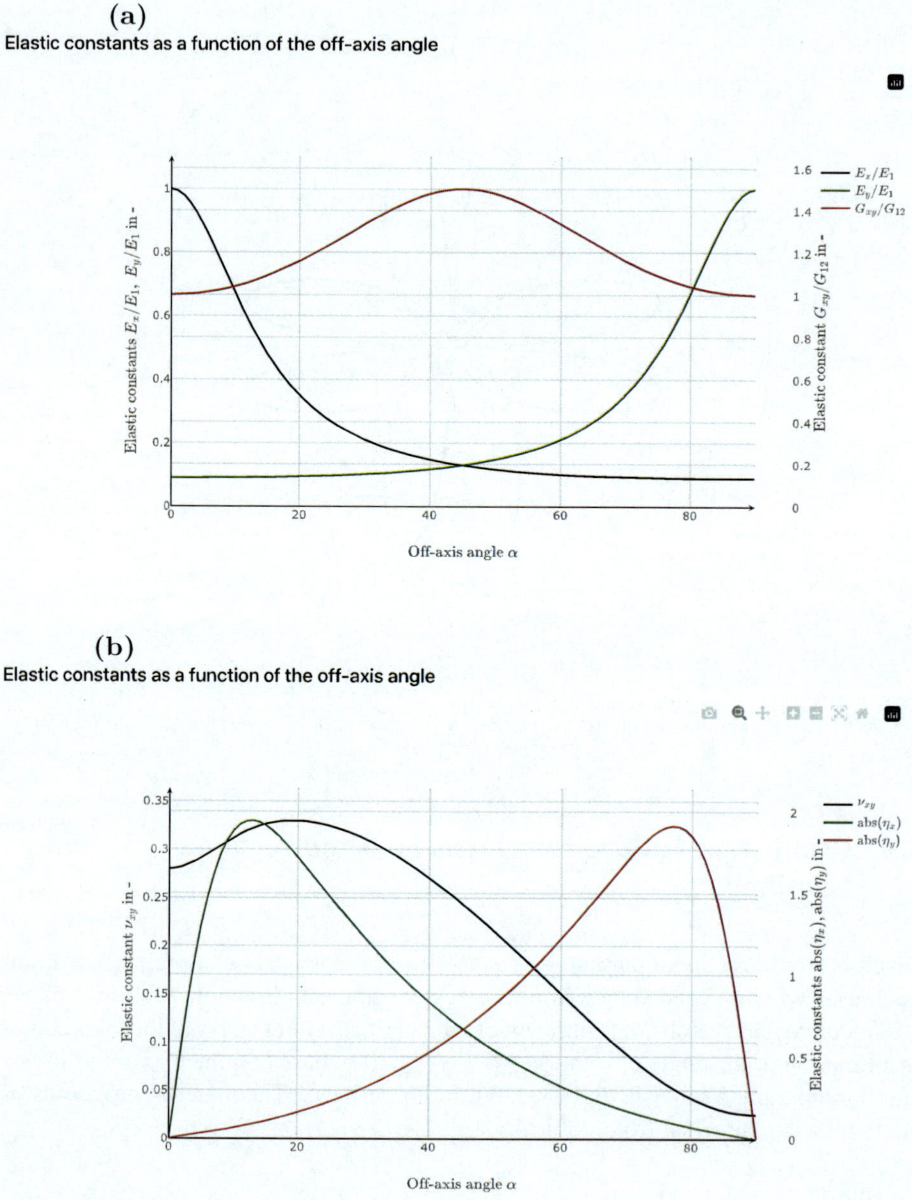

Fig. 4.61 Elastic constants as a function of the off-axis angle: **a** E_x/E_1, E_y/E_1, G_{xy}/G_{12} and **b** ν_{xy}, $-\eta_x$, $-\eta_y$

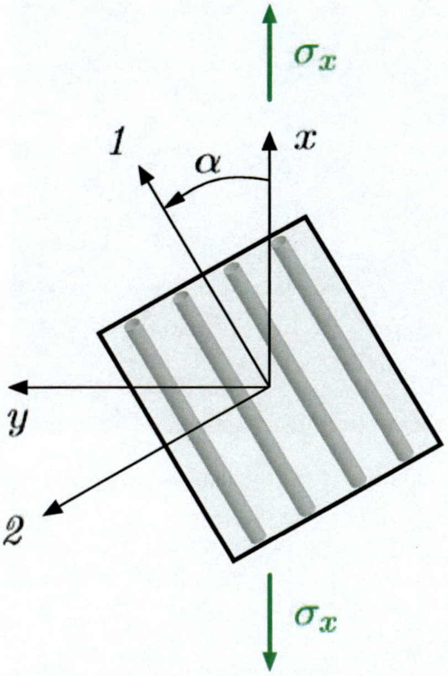

Fig. 4.62 Unidirectional lamina loaded by a single normal stress σ_x

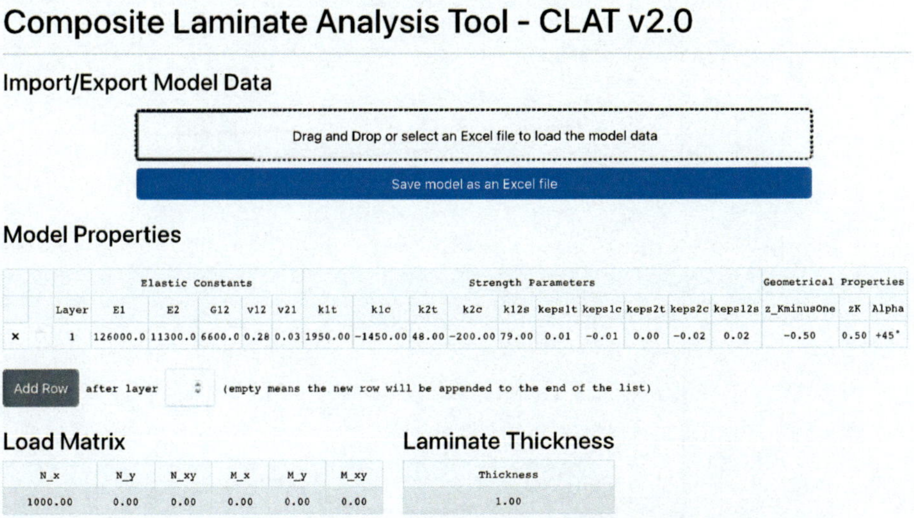

Fig. 4.63 Model definition in CLAT for the laminate shown in Fig. 4.62, pole diagrams

Fig. 4.64 Layer representation in CLAT for the laminate shown in Fig. 4.62

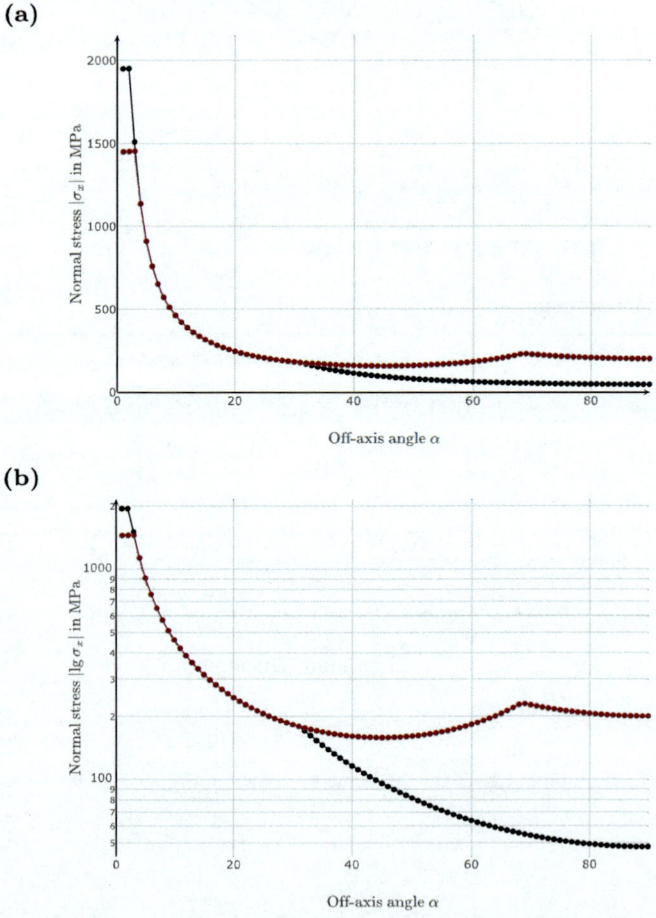

Fig. 4.65 Failure envelope for a unidirectional prepreg (see Table 4.1 for details) according to the maximum stress criterion: **a** linear division of the axes and **b** logarithmic vertical axis

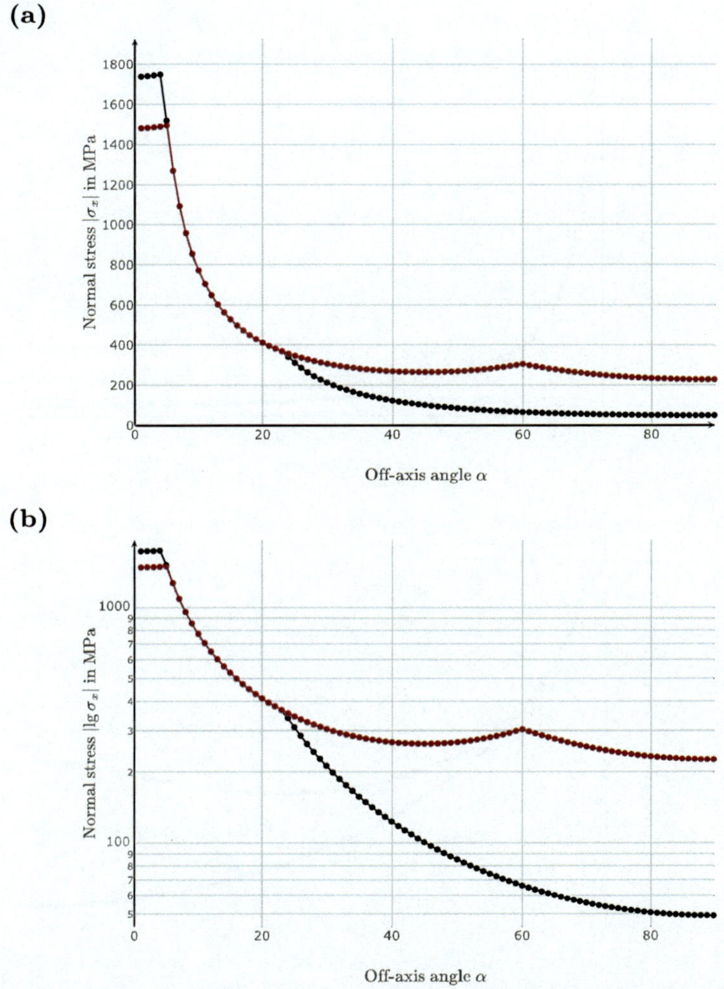

Fig. 4.66 Failure envelope for a unidirectional prepreg (see Table 4.1 for details) according to the maximum strain criterion: **a** linear division of the axes and **b** logarithmic vertical axis

The graphical representation of the unidirectional lamina in CLAT is given in Fig. 4.64. Layer number and angle of rotation are provided.

The failure envelopes according to the different failure criteria, i.e., the maximum stress, maximum strain, Tsai-Hill, and Tsai-Wu, are provided in Figs. 4.65, 4.66, 4.67, and 4.68. It can be seen that all failure criteria predict a reduced prepreg strength with increasing off-axis angle α.

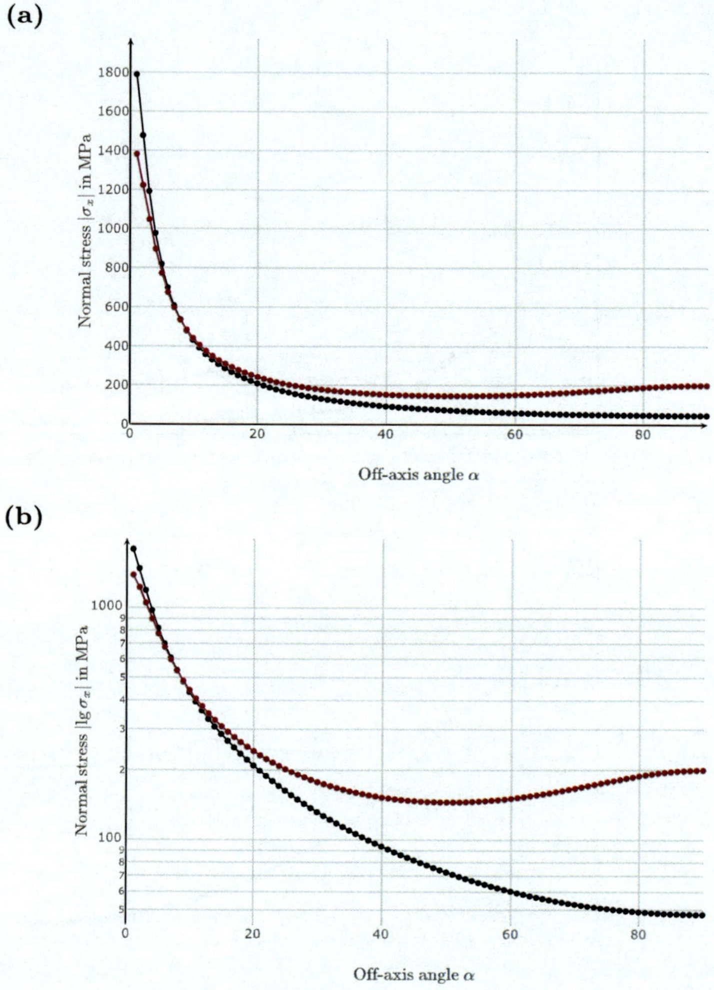

Fig. 4.67 Failure envelope for a unidirectional prepreg (see Table 4.1 for details) according to the Tsai-Hill criterion: **a** linear division of the axes and **b** logarithmic vertical axis

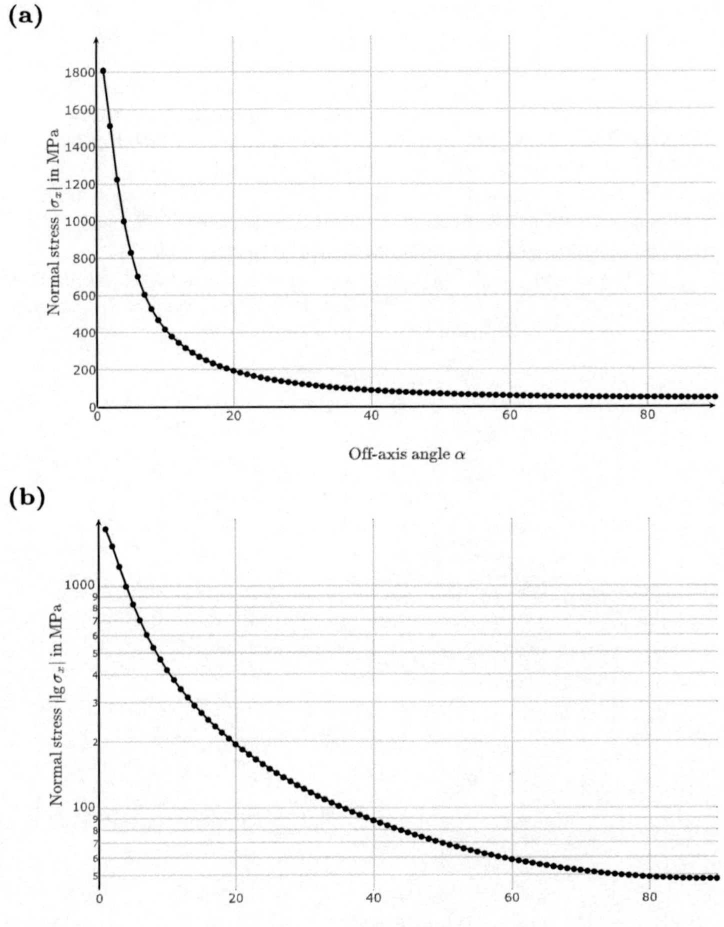

Fig. 4.68 Failure envelope for a unidirectional prepreg (see Table 4.1 for details) according to the Tsai-Wu criterion: **a** linear division of the axes and **b** logarithmic vertical axis

References

Buryachenko VA (2007) Micromechanics of heterogeneous materials. Springer, New York
Dvorak GJ (2013) Micromechanics of composite materials. Springer, New York
Herakovich CT (2012) Mechanics of composites: a historical review. Mech Res Commun 41:1–20
Kaw AK (2006) Mechanics of composite materials. CRC Press, Boca Raton
MIL-HDBK-17, 1F, (2002) Composite materials handbook-, vol 1. Polymer matrix composites guidelines for characterization of structural materials, Department of Defence, USA
Öchsner A (2021) Foundations of classical laminate theory. Springer, Cham
Soden PD, Hinton MJ, Kaddour AS (1998) Lamina properties, lay-up configurations and loading conditions for a range of fibre-reinforced composite laminates. Compos Sci Technol 58:1011–1022

Chapter 5
Source Codes

Abstract This chapter provides the commented and structured source codes of the Python files which contains all the written routines.

5.1 Main.py

To easily navigate the source code of the file `main.py`, Table 5.1 collects in chronological order all the callbacks and the corresponding line in the source code.

Table 5.1 Line numbers of the callbacks in `main.py`

Callback	Program line
addRowToDataFrame	99
unselectCells	164
styleSelectedRows	179
renderTabContent	216
updateListOfExistingModels	253
generateGraphs	275

Supplementary Information The online version contains supplementary material available at https://doi.org/10.1007/978-3-031-32975-3_5.

A. Öchsner and R. Makvandi, *A Numerical Approach to the Classical Laminate Theory of Composite Materials*, Advanced Structured Materials 189,
https://doi.org/10.1007/978-3-031-32975-3_5

main.py (main file)

```
"""
This is the main Python file of CLAT
"""
import io
import os
import fnmatch
import dash
from dash import Input, Output, State, no_update, ALL, ctx
import globalVars
from Laminates_classes import Laminate
from CLATHelpModule import *

# current directory
cwd = os.getcwd()

# find all the xlsx files in the current directory
problemFiles = sorted(fnmatch.filter(os.listdir(cwd), "*.xlsx"))

elasticProperties = {"E1": 126000.0, "E2": 11300.0, "G12": 6600.0, "v12": 0.28, "v21": 0.02511111111111112}
strengthParameters = {"k1t": 1950.0, "k1c": -1450.0, "k2t": 48.0, "k2c": -200.0, "k12s": 79.0, "keps1t": 1.38e-2,
                      "keps1c": -1.175e-2, "keps2t": 0.436e-2, "keps2c": -2.0e-2, "keps12s": 2.0e-2}
geometricalProperties = {"zKminusOne": -4.0, "zK": -3.0, "alpha": +45.0}

laminateListOfLayers = {
    0: (elasticProperties | strengthParameters | {"zKminusOne": -4.0, "zK": -3.0, "alpha": +45.0}),
    1: (elasticProperties | strengthParameters | {"zKminusOne": -3.0, "zK": -2.0, "alpha": -45.0}),
    2: (elasticProperties | strengthParameters | {"zKminusOne": -2.0, "zK": -1.0, "alpha": 0.0}),
    3: (elasticProperties | strengthParameters | {"zKminusOne": -1.0, "zK": 0.0, "alpha": +90.0}),
    4: (elasticProperties | strengthParameters | {"zKminusOne": 0.0, "zK": +1.0, "alpha": +90.0}),
    5: (elasticProperties | strengthParameters | {"zKminusOne": +1.0, "zK": +2.0, "alpha": 0.0}),
    6: (elasticProperties | strengthParameters | {"zKminusOne": +2.0, "zK": +3.0, "alpha": -45.0}),
    7: (elasticProperties | strengthParameters | {"zKminusOne": +3.0, "zK": +4.0, "alpha": +45.0})
}

loadMatrix = {
    0: {"N_x": 1000.0, "N_y": 0.0, "N_xy": 0.0, "M_x": 0.0, "M_y": 0.0, "M_xy": 0.0}
}

thickness = {
```

```python
    0: {"Thickness": 9.0}
}
thicknessDataFrame = pd.DataFrame.from_dict(thickness, orient='index', columns=["Thickness"])

globalVars.failureAnalysisFailedPlies = []
globalVars.failedPlies = []

df = pd.DataFrame.from_dict(
    laminateListOfLayers, orient='index',
    columns=['Layer', 'E1', 'E2', 'G12', 'v12', 'v21', 'k1t', 'k1c', 'k2t', 'k2c', 'k12s', 'keps1t',
             'keps1c', 'keps2t', 'keps2c', 'keps12s', 'zKminusOne', 'zK', 'alpha', 'thickness']
)
df["id"] = df.index
df['Layer'] = np.arange(0, len(df)) + 1
df['thickness'] = df['zK'] - df['zKminusOne']

loadMatrixDataFrame = pd.DataFrame.from_dict(
    loadMatrix, orient='index', columns=["N_x", "N_y", "N_xy", "M_x", "M_y", "M_xy"]
)

# initialize the Dash app with a template from Dash Bootstrap
dbc_css = "https://cdn.jsdelivr.net/gh/AnnMarieW/dash-bootstrap-templates@V1.0.6"
app = dash.Dash(__name__,
                external_stylesheets=[dbc.themes.BOOTSTRAP, dbc_css],
                suppress_callback_exceptions=True
                )
app.title = "CLAT v2.0"
app.layout = html.Div(
    dbc.Container(
        [
            dcc.Store(id="store"),
            html.H1("Composite Laminate Analysis Tool - CLAT v2.0"),
            html.Hr(),
            getLoadingOverlay(df, loadMatrixDataFrame, thicknessDataFrame),
            html.Hr(),
            getAnalysisOptionsRow(problemFiles),
            html.Hr(),
            getAnalysisTabs(),
            html.Div(id="tab-content", className="p-4")
        ]
    )
)


@app.callback(
    Output({'type': 'table', 'index': 'myTable'}, 'data'),
```

```
    Output({'type': 'table', 'index': 'myLoadMatrixTable'}, 'data'),
    Output({'type': 'table', 'index': 'thicknessMatrix'}, 'data'),
    Output('addRowAfterIndex', 'max'),
    Input('editing-rows-button', 'n_clicks'),
    Input({'type': 'table', 'index': 'myTable'}, 'data'),
    Input({'type': 'table', 'index': 'myLoadMatrixTable'}, 'data'),
    Input({'type': 'table', 'index': 'thicknessMatrix'}, 'data'),
    State('addRowAfterIndex', 'value'),
    Input('datatable-upload', 'contents'),
    State('datatable-upload', 'filename'),
    Input('dropdown-menu', 'value'),
    prevent_initial_call=True,
)
def addRowToDataFrame(n_clicks, modelPropertiesTable, loadMatrixTable, thicknessMatrixTable, index, contents, filename,
                      problemName):
    if dash.callback_context.triggered[0]['prop_id'] == 'editing-rows-button.n_clicks':
        if n_clicks > 0:
            if index is None:
                modelPropertiesTable.append(copy.copy(modelPropertiesTable[-1]))
            else:
                try:
                    modelPropertiesTable.insert(index, copy.copy(modelPropertiesTable[index - 1]))
                except Exception as e:
                    print(e)
                    print('Wrong value for the layer number!')

    if dash.callback_context.triggered[0]['prop_id'] == 'datatable-upload.contents':
        globalVars.failureAnalysisFailedPlies = []
        globalVars.failedPlies = []
        if contents is None:
            raise dash.exceptions.PreventUpdate
        [newModelProperties, newloadMatrix, newThicknessMatrix] = parse_contents(contents, filename)
        modelPropertiesTable = newModelProperties.to_dict('records')
        loadMatrixTable = newloadMatrix.to_dict('records')
        thicknessMatrixTable = newThicknessMatrix.to_dict('records')

    if dash.callback_context.triggered[0]['prop_id'] == 'dropdown-menu.value':
        if problemName is None:
            raise dash.exceptions.PreventUpdate
```

```python
        modelPropertiesTable = pd.read_excel(problemName, sheet_name="ModelProperties").to_dict('records')
        loadMatrixTable = pd.read_excel(problemName, sheet_name="LoadMatrix").to_dict('records')
        thicknessMatrixTable = pd.read_excel(problemName, sheet_name="thicknessMatrix").to_dict('records')

    updateLayerColumn(modelPropertiesTable)
    return modelPropertiesTable, loadMatrixTable, thicknessMatrixTable, len(modelPropertiesTable)


@app.callback(
    Output("download-dataframe-xlsx", "data"),
    Input("btn_xlsx", "n_clicks"),
    State({'type': 'table', 'index': "myTable"}, "data"),
    State({'type': 'table', 'index': "myLoadMatrixTable"}, "data"),
    State({'type': 'table', 'index': "thicknessMatrix"}, "data"),
    prevent_initial_call=True,
)
def saveModelAsExcel(n_clicks, data, loadMat, thicknessMat):
    output = io.BytesIO()
    excelFileWriter = pd.ExcelWriter(output, engine='xlsxwriter')
    modelPropertiesDataFrame = pd.DataFrame(data)
    loadMatDataFrame = pd.DataFrame(loadMat)
    thicknessMatDataFrame = pd.DataFrame(thicknessMat)
    modelPropertiesDataFrame.to_excel(excelFileWriter, sheet_name='ModelProperties')
    loadMatDataFrame.to_excel(excelFileWriter, sheet_name='LoadMatrix', index=False)
    thicknessMatDataFrame.to_excel(excelFileWriter, sheet_name='thicknessMatrix', index=False)
    excelFileWriter.save()
    data = output.getvalue()
    return dcc.send_bytes(data, "problem_data.xlsx")


# read the following page for more info
# https://community.plotly.com/t/how-to-reset-dash-datatable-selected-cells-varaible-to-none-from-within-a-callback/48588/4
@app.callback(
    Output({'type': 'table', 'index': ALL}, 'selected_cells'),
    Output({'type': 'table', 'index': ALL}, 'active_cell'),
    Input({'type': 'table', 'index': ALL}, 'selected_cells'),
    State({'type': 'table', 'index': ALL}, 'id'),
    prevent_initial_call=True
)
def unselectCells(cells, tables):
    table_ids = [t['index'] for t in tables]
```

```python
    trigger = ctx.triggered_id.index # only works if you have dash>=2.4.0
    new_selection = [[] if t != trigger else c for c, t in zip(cells, table_ids)]
    new_active = [None if s == [] else s[0] for s in new_selection]
    return new_selection, new_active


@app.callback(
    Output({'type': 'table', 'index': "myTable"}, "style_data_conditional"),
    Output({'type': 'table', 'index': "myTable"}, "tooltip_conditional"),
    Input({'type': 'table', 'index': "myTable"}, "derived_viewport_selected_row_ids"),
    Input({'type': 'table', 'index': "myTable"}, "style_data_conditional"),
    Input({'type': 'table', 'index': "myTable"}, "selected_rows"),
)
def styleSelectedRows(selRows, styleConditioning, rows):
    # we only want the main style conditioning to stay forever
    styleConditioning = styleConditioning[0:2]
    if selRows is None:
        return dash.no_update
    new = [
        {
            "if": {
                "filter_query": "{{id}} = {}".format(i)
            },
            'backgroundColor': '#800000',
            'color': 'lightgray',
        }
        for i in selRows]
    styleConditioning.extend(new)

    tooltip_conditional = [
        {
            'if': {
                'filter_query': "{{id}} = {}".format(i)
            },
            'type': 'markdown',
            'value': 'This layer will be **excluded** from the analyses.'
        } for i in selRows
    ]

    return styleConditioning, tooltip_conditional


@app.callback(
    Output("tab-content", "children"),
    Input("tabs", "active_tab"),
```

```
    Input("store", "data"),
    Input("failureAnalysis", "value"),
    Input("plyByPlyAnalysisCheckBox", "value"),
    Input("failureEnvelopesCheckBox", "value")
)
def renderTabContent(active_tab, data, failureAnalysis, plyByPlyAnalysis, failureEnvelopes):
    """
    This callback takes the 'active_tab' property as input, as well as the
    stored graphs, and renders the tab content depending on what the value of
    'active_tab' is.
    """

    if active_tab and data is not None:
        if active_tab == "tabSketch":
            return getCustomizedGraph(data["sketch"], {"fileName" : "sketch"})
        elif active_tab == "tabStresses":
            return data["stressAnalysis"]
        elif active_tab == "tabStrains":
            return data["strainAnalysis"]
        elif (active_tab == "tabFailureAnalysis") and ("Failure Analysis" not in failureAnalysis):
            return dbc.Alert("Failure Analysis is not active.",color="info")
        elif (active_tab == "tabFailureAnalysis") and ("Failure Analysis" in failureAnalysis):
            return data["failureAnalysis"]
        elif (active_tab == "tabPlyByPly") and ("Ply-By-Ply Analysis" not in plyByPlyAnalysis):
            return dbc.Alert("Ply-By-Ply Analysis is not active.", color="info")
        elif (active_tab == "tabPlyByPly") and ("Ply-By-Ply Analysis" in plyByPlyAnalysis):
            return data["plyByPly"]
        elif active_tab == "tabUnidirectionalLaminate":
            return data["unidirectionalLaminate"]
        elif (active_tab == "tabFailureEnvelopes") and ("Failure Envelopes" not in failureEnvelopes):
            return dbc.Alert("Failure Envelopes analysis is not active.", color="info")
        elif (active_tab == "tabFailureEnvelopes") and ("Failure Envelopes" in failureEnvelopes):
            return data["failureEnvelopes"]

    return ""


@app.callback(
    Output("dropdown-menu", "options"),
    Input("dropdownParent", 'n_clicks'),
    prevent_initial_call=True,
)
def updateListOfExistingModels(n):
```

```
    # current directory
    # cwd = os.getcwd()
    # find all the Excel files in the current directory
    modelFiles = fnmatch.filter(os.listdir(cwd), "*.xlsx")
    return sorted(modelFiles)


@app.callback(Output("store", "data"),
              Output("loading-output", "children"),
              Output({'type': 'table', 'index': 'myTable'},'selected_rows'),
              Input("button", "n_clicks"),
              Input({'type': 'table', 'index': "myTable"}, "data"),
              Input({'type': 'table', 'index': "myLoadMatrixTable"},"data"),
              Input({'type': 'table', 'index': "thicknessMatrix"},"data"),
              Input({'type': 'table', 'index': "myTable"},"derived_virtual_selected_rows"),
              State("automaticUpdate", "value"),
              Input("failureAnalysis", "value"),
              Input("plyByPlyAnalysisCheckBox", "value"),
              Input("failureEnvelopesCheckBox", "value"),
              prevent_initial_call=True,
              )
def generateGraphs(n, rows, loadMat, thicknessMat, userSelectedFailedPlies, automaticUpdate, failureAnalysis,
                   plyByPlyAnalysis, failureEnvelopesAnalysis):
    globalVars.failedPlies = globalVars.failureAnalysisFailedPlies+ userSelectedFailedPlies
    changed_id = [p['prop_id'] for p in dash.callback_context.triggered][0]
    if OR("Automatic Update" in automaticUpdate, "button" inchanged_id):
        laminateSketch = getLaminateSketch(rows, userSelectedFailedPlies)
        laminateObject = Laminate({})

        dataFrame = pd.DataFrame(rows)
        dataFrame = dataFrame.drop(['Layer'], axis=1)
        laminateObject.addLaminae(dataFrame.to_dict('index'))

        loadMatDataFrame = pd.DataFrame(loadMat)
        laminateObject.updateLoadMatrix(loadMatDataFrame.to_numpy().transpose())

        laminateObject.setThickness(pd.DataFrame(thicknessMat)["Thickness"])

        failureAnalysisContainer = {}
        if "Failure Analysis" in failureAnalysis:
```

```
            failureAnalysisContainer = getFailureAnalysisContainerlaminateObject)

        plyByPlyContainer = {}
        if "Ply-By-Ply Analysis" in plyByPlyAnalysis:
            backup = globalVars.failureAnalysisFailedPlies
            backup2 = globalVars.failedPlies
            globalVars.failureAnalysisFailedPlies = []
            globalVars.failedPlies = []
            plyByPlyContainer = getPlyByPlyAnalysis(laminateObject, userSelectedFailedPlies)
            globalVars.failedPlies = backup2
            globalVars.failureAnalysisFailedPlies = backup

        failureEnvelopesContainer = {}
        if "Failure Envelopes" in failureEnvelopesAnalysis:
            failureEnvelopesContainer = getFailureEnvelopesContainer(laminateObject)

        return {"sketch": laminateSketch,
                "stressAnalysis": getStressAnalysisContainer(laminateObject),
                "strainAnalysis": getStrainAnalysisContainer(laminateObject),
                "unidirectionalLaminate": getUnidirectionalLaminateContainer(laminateObject),
                "failureEnvelopes": failureEnvelopesContainer,
                "failureAnalysis": failureAnalysisContainer,
                "plyByPly": plyByPlyContainer,
                }, no_update, userSelectedFailedPlies
    else:
        raise dash.exceptions.PreventUpdate


if __name__ == "__main__":
    app.run_server(debug=True)
```

5.2 Laminates_classes.py

To easily navigate the source code of the file `Laminates_classes.py`, Table 5.2 collects in chronological order all the classes methods and the corresponding line in the source code.

Table 5.2 Line numbers of the classes methods in `Laminates_classes.py`

Classes method	Program line
getLayerNo	16
getTransformedElasticityMatrix	19
getTransformedComplianceMatrix	22
getElasticityMatrix	25
setProperties	42
getProperties	45
getComplianceMatrix	48
getStressTransformationMatrix	51
getStrainTransformationMatrix	63
getZK	76
getZKminusOne	79
transformElasticityMatrix	82
transformComplianceMatrix	90
__concatenateGeneralizedElasticitySubmatrices	111
__getBendingSubmatrix	117
__getCouplingSubmatrix	129
__getExtensionalSubmatrix	142
__updateNumberOfLaminateLayers	153
addLaminae	156
createLamina	161
findLayerIndex	164
getEpsMatrix	172
getGeneralizedComplianceMatrix	175
getGeneralizedElasticityMatrix	178
getGeneralizedStrains	184
getKappaMatrix	187
getLaminate	190
getLaminateLayer	193
getLaminateLayers	196
getLoadMatrix	199
getNumberOfLaminateLayers	202
getStrainsIn12	205
getStrainsInXY	217
getStressesIn12	226
getStressesInXY	236
getThickness	247
removeLamina	250
setThickness	254
updateLamina	257
updateLaminate	262
updateLoadMatrix	268

Laminates_classes.py (laminates classes)

```python
import numpy as np
import globalVars
from decimal import *


class Lamina(object):
    def __init__(self, properties, k):
        self.Properties = properties
        self.layerNo = k
        # self.strengthParameters = propertiesTuple[1]
        # self.geometricalProperties = propertiesTuple[2]

        self.elasticityMatrix = self.getElasticityMatrix()
        self.transformedElasticityMatrix = self.getTransformedElasticityMatrix()

    def getLayerNo(self):
        return self.layerNo

    def getTransformedElasticityMatrix(self):
        return self.transformElasticityMatrix(self.elasticityMatrix)

    def getTransformedComplianceMatrix(self):
        return self.transformComplianceMatrix(self.getComplianceMatrix())

    def getElasticityMatrix(self):
        # elastic constants
        E1 = self.Properties["E1"]
        E2 = self.Properties["E2"]
        v12 = self.Properties["v12"]
        v21 = self.Properties["v21"]
        G12 = self.Properties["G12"]

        C11 = E1 / (1.0 - v12*v21)
        C22 = E2 / (1.0 - v12*v21)
        C12 = E1*v21 / (1.0 - v12*v21)
        C21 = E2*v12 / (1.0 - v12*v21)
        elasticityMatrix = np.array([[C11, C12, 0.0],
                                     [C21, C22, 0.0],
                                     [0.0, 0.0, G12]])
        return elasticityMatrix

    def setProperties(self, focusedProperty, newValue):
        self.Properties[focusedProperty] = newValue

    def getProperties(self):
        return self.Properties

    def getComplianceMatrix(self):
        return np.linalg.inv(self.getElasticityMatrix())

    def getStressTransformationMatrix(self):
        # geometric properties
```

```python
        alpha = self.Properties["alpha"]

        cosAlpha = np.cos(alpha * np.pi / 180.0)
        sinAlpha = np.sin(alpha * np.pi / 180.0)

        transformationMatrix = np.array([[cosAlpha ** 2, sinAlpha ** 2, 2.0 * sinAlpha * cosAlpha],
                                         [sinAlpha ** 2, cosAlpha ** 2, -2.0 * sinAlpha * cosAlpha],
                                         [-sinAlpha * cosAlpha, sinAlpha * cosAlpha, cosAlpha ** 2 - sinAlpha ** 2]])
        return transformationMatrix

    def getStrainTransformationMatrix(self):
        # geometric properties
        alpha = self.Properties["alpha"]

        cosAlpha = np.cos(alpha * np.pi / 180.0)
        sinAlpha = np.sin(alpha * np.pi / 180.0)

        transformationMatrix = np.array([[cosAlpha ** 2, sinAlpha ** 2, sinAlpha * cosAlpha],
                                         [sinAlpha ** 2, cosAlpha ** 2, -sinAlpha * cosAlpha],
                                         [-2 * sinAlpha * cosAlpha, 2 * sinAlpha * cosAlpha,
                                          cosAlpha ** 2 - sinAlpha ** 2]])
        return transformationMatrix

    def getZK(self):
        return self.Properties["zK"]

    def getZKminusOne(self):
        return self.Properties["zKminusOne"]

    def transformElasticityMatrix(self, originalMatrix):
        stressTransformationMatrix = self.getStressTransformationMatrix()
        strainTransformationMatrix = self.getStrainTransformationMatrix()

        theTransformedMatrix = originalMatrix.dot(strainTransformationMatrix)
        theTransformedMatrix = np.linalg.inv(stressTransformationMatrix).dot(theTransformedMatrix)
        return theTransformedMatrix

    def transformComplianceMatrix(self, originalMatrix):
        stressTransformationMatrix = self.getStressTransformationMatrix()
        strainTransformationMatrix = self.getStrainTransformationMatrix()

```

```
        theTransformedMatrix = originalMatrix.dot(stressTransformationMatrix)
        theTransformedMatrix = np.linalg.inv(strainTransformationMatrix).dot(theTransformedMatrix)
        return theTransformedMatrix


class Laminate(object):
    def __init__(self, laminateLayers):
        self.laminateLayers = laminateLayers
        self.NumberOfLaminateLayers = len(laminateLayers)
        self.laminae = []
        self.updateLaminate()
        self.loadMatrix = np.zeros((6, 1))
        self.thickness = 9.0

    def __iter__(self):
        return (lamina for lamina in self.laminae)

    def __concatenateGeneralizedElasticitySubmatrices(self, A, B, D):
        cStarFirstRow = np.concatenate((A, B), axis=1)
        cStarSecondRow = np.concatenate((B, D), axis=1)
        cStar = np.concatenate((cStarFirstRow, cStarSecondRow), axis=0)
        return cStar

    def __getBendingSubmatrix(self):
        D = np.zeros((3, 3))
        for k in np.arange(len(self.laminae)):
            if globalVars.isNotFailed(k):
                currentLamina = self.laminae[k]
                transformedElasticityMatrix = currentLamina.getTransformedElasticityMatrix()
                zK = currentLamina.getZK()
                zKminusOne = currentLamina.getZKminusOne()
                D = D + transformedElasticityMatrix*(zK**3 - zKminusOne**3)
        D = (1.0/3.0)*D
        return D

    def __getCouplingSubmatrix(self):
        B = np.zeros((3, 3))
        for k in np.arange(len(self.laminae)):
            if globalVars.isNotFailed(k):
                currentLamina = self.laminae[k]
                transformedElasticityMatrix = currentLamina.getTransformedElasticityMatrix()
                zK = currentLamina.getZK()
                zKminusOne = currentLamina.getZKminusOne()
                B = B + transformedElasticityMatrix*(zK**2 - zKminusOne**2)
        B = (1.0/2.0)*B

        return B

    def __getExtensionalSubmatrix(self):
        A = np.zeros((3, 3))
```

```
        for k in np.arange(len(self.laminae)):
            if globalVars.isNotFailed(k):
                currentLamina = self.laminae[k]
                transformedElasticityMatrix = currentLamina.
                    ↪ getTransformedElasticityMatrix()
                zK = currentLamina.getZK()
                zKminusOne = currentLamina.getZKminusOne()
                A = A + transformedElasticityMatrix*(zK - zKminusOne)
        return A

    def __updateNumberOfLaminateLayers(self):
        self.NumberOfLaminateLayers = len(self.laminateLayers)

    def addLaminae(self, newLaminae):
        self.laminateLayers.update(newLaminae)
        self.__updateNumberOfLaminateLayers()
        self.updateLaminate()

    def createLamina(self, laminateData, k):
        return Lamina(laminateData, k)

    def findLayerIndex(self, zCoordinate):
        for k in np.arange(len(self.laminae)):
            currentLamina = self.laminae[k]
            zK = currentLamina.getZK()
            zKminusOne = currentLamina.getZKminusOne()
            if zKminusOne <= zCoordinate <= zK:
                return k

    def getEpsMatrix(self):
        return self.getGeneralizedStrains()[0:3]

    def getGeneralizedComplianceMatrix(self):
        return np.linalg.inv(self.getGeneralizedElasticityMatrix())

    def getGeneralizedElasticityMatrix(self):
        A = self.__getExtensionalSubmatrix()
        B = self.__getCouplingSubmatrix()
        D = self.__getBendingSubmatrix()
        return self.__concatenateGeneralizedElasticitySubmatrices(A, B,
            ↪ D)

    def getGeneralizedStrains(self):
        return self.getGeneralizedComplianceMatrix().dot(self.loadMatrix
            ↪ )

    def getKappaMatrix(self):
        return self.getGeneralizedStrains()[3:]

    def getLaminate(self):
        return self.laminae

    def getLaminateLayer(self, layerNumber):
        return self.laminae[layerNumber]

    def getLaminateLayers(self):
```

```
        return self.laminateLayers

    def getLoadMatrix(self):
        return self.loadMatrix

    def getNumberOfLaminateLayers(self):
        return self.NumberOfLaminateLayers

    def getStrainsIn12(self, zCoordinate):
        layerIndex = self.findLayerIndex(zCoordinate)
        if globalVars.isNotFailed(layerIndex):
            lamina = self.laminae[layerIndex]
            # complianceMatrix = lamina.getComplianceMatrix()
            # sigmaMatrix = self.getStressesIn12(zCoordinate)
            strainsXY = self.getStrainsInXY(zCoordinate)
            transformationMatrix = lamina.getStrainTransformationMatrix()
            return transformationMatrix.dot(strainsXY)
        else:
            return np.array([[0.0], [0.0], [0.0]])

    def getStrainsInXY(self, zCoordinate):
        layerIndex = self.findLayerIndex(zCoordinate)
        if globalVars.isNotFailed(layerIndex):
            epsMatrix = self.getEpsMatrix()
            kappaMatrix = self.getKappaMatrix()
            return epsMatrix + zCoordinate*kappaMatrix
        else:
            return np.array([[0.0], [0.0], [0.0]])

    def getStressesIn12(self, zCoordinate):
        layerIndex = self.findLayerIndex(zCoordinate)
        if globalVars.isNotFailed(layerIndex):
            lamina = self.laminae[layerIndex]
            sigmaXY = self.getStressesInXY(zCoordinate)
            transformationMatrix = lamina.getStressTransformationMatrix()
            return transformationMatrix.dot(sigmaXY)
        else:
            return np.array([[0.0], [0.0], [0.0]])

    def getStressesInXY(self, zCoordinate):
        layerIndex = self.findLayerIndex(zCoordinate)
        if globalVars.isNotFailed(layerIndex):
            lamina = self.laminae[layerIndex]
            transformedElasticiyMatrix = lamina.getTransformedElasticityMatrix()
            epsMatrix = self.getEpsMatrix()
            kappaMatrix = self.getKappaMatrix()
            return transformedElasticiyMatrix.dot(epsMatrix + zCoordinate*kappaMatrix)
        else:
            return np.array([[0.0], [0.0], [0.0]])

    def getThickness(self):
        return self.thickness
```

```

    def removeLamina(self, laminaItemNumber):
        self.laminateLayers.pop(laminaItemNumber)
        self.__updateNumberOfLaminateLayers()

    def setThickness(self, newValue):
        self.thickness = newValue

    def updateLamina(self, newLaminae):
        self.laminateLayers.update(newLaminae)
        self.__updateNumberOfLaminateLayers()
        self.updateLaminate()

    def updateLaminate(self):
        self.laminae = []
        for k in self.laminateLayers.keys():
            # print(self.laminateLayers[k])
            self.laminae.append(self.createLamina(self.laminateLayers[k
                ↪ ], k))

    def updateLoadMatrix(self, newLoadMatrix):
        self.loadMatrix = newLoadMatrix
```

5.3 GlobalVars.py

To easily navigate the source code of the file `globalVars.py`, Table 5.3 collects in chronological order the list of functions and the corresponding line in the source code.

globalVars.py (global variables)

```
1 failureAnalysisFailedPlies = []
2 failedPlies = []
3
4 def isNotFailed(plyIndex):
5     if plyIndex in failedPlies:
6         return False
7     else:
8         return True
```

Table 5.3 Line numbers of the list of functions in `globalVars.py`

Function	Program line
isNotFailed	4

5.4 CLATHelpModule.py

To easily navigate the source code of the file `CLATHelpModule.py`, Table 5.4 collects in chronological order the list of functions and the corresponding line in the source code.

Table 5.4 Line numbers of the list of functions in `CLATHelpModule.py`

Function	Program line
OR	20
getAccordionDataTable	27
getAnalysisOptionColumn	78
getAnalysisOptionsRow	89
getAnalysisRegenerateGraphsColumn	121
getAnalysisTabs	133
getCustomizedGraph	149
getDataTable	170
getDataTableManipulationRow	249
getDataTableObject	269
getDataTableWithTooltip	311
getDownloadButtons	333
getFailureAnalysisContainer	340
getFailureAnalysisDirectionColumn	421
getFailureAnalysisLaminaeColumn	431
getFailureAnalysisMaxStrainColumn	441
getFailureAnalysisMaxStressColumn	476
getFailureAnalysisPositionColumn	510
getFailureAnalysisTsaiHillColumn	521
getFailureAnalysisTsaiWuColumn	543
getFailureEnvelopesContainer	564
getFailureEnvelopesRMaximumStrainCriteria	901
getFailureEnvelopesRMaximumStressCriteria	953
getFailureModeTable	1004
getLaminateSketch	1063
getLaminateThicknessTable	1116
getLaminateThicknessTableWithTooltip	1150
getLaminateThicknessTableWithTooltipColumn	1168
getLayerColumn	1175
getLoadExistingModelsModule	1184
getLoadExistingModelsModuleWithTooltip	1198

(continued)

Table 5.4 (continued)

Function	Program line
getLoadExistingModelsModuleWithTooltipColumn	1217
getLoadTable	1224
getLoadTableColumn	1263
getLoadingOverlay	1270
getPlyByPlyAnalysis	1299
getPositionArray	1485
getPositionColumn	1494
getRTsaiHill	1503
getRTsaiWu	1536
getStrain12DataFrame	1561
getStrain12Values	1572
getStrainAnalysisContainer	1584
getStrainPlots	1726
getStrainXYValues	1788
getStrengthRatioDataFrame	1800
getStrengthRatioValues	1810
getStress12DataFrame	1823
getStress12Values	1834
getStressAnalysisContainer	1847
getStressPlots	1993
getStressXYValues	2054
getTableDataFrame	2066
getUnidirectionalLaminateContainer	2074
getUploadButtons	2410
getYValues	2427
makeLinePlot	2437
makePlot	2546
makePolarDiagram	2618
makeStressStrainPlot	2655
parse_contents	2725
updateLayerColumn	2735

CLATHelpModule.py (help module)

```
from dash import dcc, html, dash_table
from dash.dash_table.Format import Format, Scheme, Symbol, Sign
from plotly.subplots import make_subplots
from sympy import Symbol as SympySymbol
from sympy.solvers import solve
import base64
import copy
```

```

import dash_bootstrap_components as dbc
import dash_mantine_components as dmc
import globalVars


import io
import numpy as np
import pandas as pd
import plotly.express as px
import plotly.graph_objs as go

def OR(a, b):
    if a == 1 or b == 1:
        return True
    else:
        return False


def getAccordionDataTable(plotData, dataTableValues):
    columns = []
    for subplot in plotData.keys():
        if plotData[subplot]["title"] in dataTableValues[0].keys():
            columns.append(
                dict(id=plotData[subplot]['title'],
                     name=[plotData[subplot]['title']],
                     type='numeric',
                     format=Format(precision=plotData[subplot]['precision
                         ↪ '] if 'precision' in plotData[subplot] else
                         ↪ 3,
                                   scheme=Scheme.fixed))
            )
    return dbc.Accordion(
        [
            dbc.AccordionItem(
                dash_table.DataTable(
                    data=dataTableValues,
                    columns=columns,
                    style_header={
                        'textAlign': 'center'
                    },
                    style_cell={
                        'textAlign': 'center'
                    },
                    style_data={
                        'color': 'black',
                        'backgroundColor': 'white'
                    },
                    style_data_conditional=[
                        {
                            'if': {'row_index': 'odd'},
                            'backgroundColor': 'rgb(220,␣220,␣220)',
                        },
                        {
                            "if": {"state": "selected"}, # 'active' | '
                                ↪ selected'
```

```
                            "backgroundColor": "#c8f4ff",
                            "border": "1px␣solid␣darkblue",
                        },
                    ],
                    export_format='xlsx',
                    export_headers='ids',
                    id='tbl',
                    page_current=0,
                    page_size=10
                ),
                title="Data␣Table",
            )
        ],
        start_collapsed=True,
    )


def getAnalysisOptionColumn(title, idValue, value):
    return dmc.Col(
        dcc.Checklist(
            options=[title],
            id=idValue,
            value=[value]
        ),
        span=2
    )


def getAnalysisOptionsRow(problemFiles):
    return dmc.Grid(
        children=[
            getLoadExistingModelsModuleWithTooltipColumn(problemFiles),
            getAnalysisRegenerateGraphsColumn(),
            getAnalysisOptionColumn(
                title='Automatic Update',
                idValue='automaticUpdate',
                value='Automatic Update'
            ),
            getAnalysisOptionColumn(
                title='Failure Analysis',
                idValue='failureAnalysis',
                value=None
            ),
            getAnalysisOptionColumn(
                title='Ply-By-Ply Analysis',
                idValue='plyByPlyAnalysisCheckBox',
                value=None
            ),
            getAnalysisOptionColumn(
                title='Failure Envelopes',
                idValue='failureEnvelopesCheckBox',
                value=None
            )
        ],
        gutter=0,
        justify="center",
```

```python
        align="center",
    )


def getAnalysisRegenerateGraphsColumn():
    return dmc.Col(
        dbc.Button(
            "Regenerate graphs",
            color="primary",
            id="button",
            className="d-flex align-items-center"
        ),
        span=2
    )


def getAnalysisTabs():
    return dbc.Tabs(
        [
            dbc.Tab(label="Sketch", tab_id="tabSketch"),
            dbc.Tab(label="Stresses", tab_id="tabStresses"),
            dbc.Tab(label="Strains", tab_id="tabStrains"),
            dbc.Tab(label="Failure Analysis", tab_id="tabFailureAnalysis"),
            dbc.Tab(label="Ply-by-Ply", tab_id="tabPlyByPly"),
            dbc.Tab(label="Unidirectional Laminate", tab_id="tabUnidirectionalLaminate"),
            dbc.Tab(label="Failure Envelopes", tab_id="tabFailureEnvelopes"),
        ],
        id="tabs",
        active_tab="tabSketch",
    )


def getCustomizedGraph(figure, fileData):
    fileName = fileData["fileName"] if "fileName" in fileData else"plot"
    fileFormat = fileData["fileFormat"] if "fileFormat" in fileData else "svg"
    imageScale = fileData["imageScale"] if "imageScale" in fileData else 1.5
    return dcc.Graph(
        figure=figure,
        mathjax=True,
        config={
            'displayModeBar': True,
            "toImageButtonOptions": {
                "filename": fileName,
                "format": fileFormat,
                # "height": imageHeight,
                "scale": imageScale
            },
            "showTips": True,
        }

    )

```

```python

def getDataTable(dataFrame):
    return dash_table.DataTable(
        data=dataFrame.to_dict('records'),
        columns=[
            dict(id='Layer', name=['', 'Layer'], type='numeric', format=Format()),
            dict(id='E1', name=['Elastic Constants', 'E1'], type='numeric',
                format=Format(precision=1, scheme=Scheme.fixed)),
            dict(id='E2', name=['Elastic Constants', 'E2'], type='numeric',
                format=Format(precision=1, scheme=Scheme.fixed)),
            dict(id='G12', name=['Elastic Constants', 'G12'], type='numeric',
                format=Format(precision=1, scheme=Scheme.fixed)),
            dict(id='v12', name=['Elastic Constants', 'v12'], type='numeric',
                format=Format(precision=2, scheme=Scheme.fixed)),
            dict(id='v21', name=['Elastic Constants', 'v21'], type='numeric',
                format=Format(precision=2, scheme=Scheme.fixed)),
            dict(id='k1t', name=['Strength Parameters', 'k1t'], type='numeric',
                format=Format(precision=2, scheme=Scheme.fixed)),
            dict(id='k1c', name=['Strength Parameters', 'k1c'], type='numeric',
                format=Format(precision=2, scheme=Scheme.fixed)),
            dict(id='k2t', name=['Strength Parameters', 'k2t'], type='numeric',
                format=Format(precision=2, scheme=Scheme.fixed)),
            dict(id='k2c', name=['Strength Parameters', 'k2c'], type='numeric',
                format=Format(precision=2,scheme=Scheme.fixed)),
            dict(id='k12s', name=['Strength Parameters', 'k12s'], type='numeric',
                format=Format(precision=2, scheme=Scheme.fixed)),
            dict(id='keps1t', name=['Strength Parameters', 'keps1t'], type='numeric',
                format=Format(precision=2, scheme=Scheme.fixed)),
            dict(id='keps1c', name=['Strength Parameters', 'keps1c'], type='numeric',
                format=Format(precision=2, scheme=Scheme.fixed)),
            dict(id='keps2t', name=['Strength Parameters', 'keps2t'], type='numeric',
                format=Format(precision=2, scheme=Scheme.fixed)),
            dict(id='keps2c', name=['Strength Parameters', 'keps2c'], type='numeric',
                format=Format(precision=2, scheme=Scheme.fixed)),
            dict(id='keps12s', name=['Strength Parameters', 'keps12s'], type='numeric',
                format=Format(precision=2, scheme=Scheme.fixed)),
            dict(id='zKminusOne', name=['Geometrical Properties', 'z_KminusOne'], type='numeric',
                format=Format(precision=2, scheme=Scheme.fixed)),
            dict(id='zK', name=['Geometrical Properties', 'zK'], type='numeric',
                format=Format(precision=2, scheme=Scheme.fixed)),
            dict(id='alpha', name=['Geometrical Properties', 'Alpha'], type='numeric',
                format=Format(precision=0, scheme=Scheme.fixed, symbol=Symbol.yes, symbol_suffix=u'Ëš',
```

```
                                sign=Sign.positive)),
        ],
        fill_width=True,
        id={'type': 'table', 'index': 'myTable'},
        merge_duplicate_headers=True,
        style_header={
            'textAlign': 'center'
        },
        style_cell={
            'textAlign': 'center',
            'height': '32px'
        },
        style_data={
            'color': 'black',
            'backgroundColor': 'white'
        },
        style_data_conditional=[
            {
                'if': {'row_index': 'odd'},
                'backgroundColor': 'rgb(220, 220, 220)',
            },
            {
                "if": {"state": "selected"}, # 'active' | 'selected'
                "backgroundColor": "#c8f4ff",
                "border": "1px solid darkblue",
            },
        ],
        editable=True,
        row_deletable=True,
        row_selectable='multi',
        page_action='native',
        page_current=0,
        page_size=10,
        tooltip_delay=0,
        tooltip_duration=None
    )


def getDataTableManipulationRow(dataFrame):
    return dbc.Row(html.Div(
        [
            dbc.Button('Add Row', id='editing-rows-button', n_clicks=0,
                ↪ color="secondary",
                       className="d-flex align-items-center"),
            html.Plaintext(" after layer "),
            dbc.Input(
                id='addRowAfterIndex',
                type='number',
                value=None,
                style={'textAlign': 'center', 'width': '60px'},
                min=1,
                max=len(dataFrame.to_dict('records'))
            ),
            html.Plaintext(" (empty means the new row will be ↪ appended
                 to the end of the list)")
        ],
```

```
        style={'display': 'flex', 'flex-direction': 'row'}
    ))


def getDataTableObject(tableData, dataTableValues):
    columns = []
    for column in tableData.keys():
        columns.append(
            dict(id=tableData[column]['title'],
                 name=[tableData[column]['title']],
                 type=tableData[column]['type'] if 'type' in tableData[column] else 'numeric',
                 format=Format(precision=tableData[column]['precision'] if 'precision' in tableData[column] else 3,
                               scheme=Scheme.fixed))
        )
    return dash_table.DataTable(
        data=dataTableValues,
        columns=columns,
        style_header={
            'textAlign': 'center'
        },
        style_cell={
            'textAlign': 'center'
        },
        style_data={
            'color': 'black',
            'backgroundColor': 'white'
        },
        style_data_conditional=[
            {
                'if': {'row_index': 'odd'},
                'backgroundColor': 'rgb(220, 220, 220)',
            },
            {
                "if": {"state": "selected"}, # 'active' | 'selected'
                "backgroundColor": "#c8f4ff",
                "border": "1px solid darkblue",
            },
        ],
        export_format='xlsx',
        export_headers='ids',
        id='tbl',
        page_current=0,
        page_size=10
    )


def getDataTableWithTooltip(dataFrame):
    return dmc.Tooltip(
        multiline=True,
        width="auto",
        withArrow=True,
        transition="fade",
        position="bottom",
        # placement="center",
```

```
            color="gray",
            transitionDuration=500,
            label=[
                html.Li(
                    'It is possible to delete individual layers by clicking on the "x" icon on the left side of the '
                    'corresponding table row.'),
                html.Li(
                    'It is possible to supress/pre-fail layers from the analysis by selecting the "\u25A1" icon of the '
                    'corresponding layers.'),
            ],
            children=getDataTable(dataFrame)
        )


def getDownloadButtons():
    return [
        dbc.Button("Save model as an Excel file",id="btn_xlsx", style={"width": "800px"}),
        dcc.Download(id="download-dataframe-xlsx"),
    ]


def getFailureAnalysisContainer(laminateObject):
    strengthRatioDataFrame = getStrengthRatioDataFrame(laminateObject)
    return dbc.Container(
        dbc.Row(
            dash_table.DataTable(
                strengthRatioDataFrame,
                columns=[
                    dict(id='Layer no.',
                         name=['', 'Layer no.'],
                         type='numeric',
                         format=Format()),
                    dict(id='Position',
                         name=['', 'Position'],
                         type='text',
                         format=Format()),
                    dict(id='Direction',
                         name=['', 'Direction'],
                         type='numeric',
                         format=Format(precision=0,
                                       scheme=Scheme.fixed)),
                    dict(id='Stress Max',
                         name=['Strength Ration R', 'Stress Max'],
                         type='numeric',
                         format=Format(precision=3,
                                       scheme=Scheme.fixed)),
                    dict(id='Strain Max',
                         name=['Strength Ration R', 'Strain Max'],
                         type='numeric',
                         format=Format(precision=3,
                                       scheme=Scheme.fixed)),
                    dict(id='Tsai-Hill',
                         name=['Strength Ration R', 'Tsai-Hill'],
```

```
                         type='numeric',
                         format=Format(precision=3,
                                       scheme=Scheme.fixed)),
                    dict(id='Tsai-Wu',
                         name=['Strength Ration R', 'Tsai-Wu'],
                         type='numeric',
                         format=Format(precision=3,
                                       scheme=Scheme.fixed))
                ],
                merge_duplicate_headers=True,
                style_header={
                    'textAlign': 'center'
                },
                style_cell={
                    'textAlign': 'center'
                },
                style_data={
                    'color': 'black',
                    'backgroundColor': 'white'
                },
                style_data_conditional=[
                    {
                        'if': {
                            'filter_query': '{Tsai-Hill} < 1.0 && {Tsai-Hill} > 0.0',
                        },
                        'backgroundColor': 'red',
                        'color': 'white'
                    },
                    {
                        'if': {
                            'filter_query': '{Tsai-Wu} < 1.0 && {Tsai-Wu} > 0.0',
                        },
                        'backgroundColor': 'red',
                        'color': 'white'
                    },
                    {
                        "if": {"state": "selected"},
                        "backgroundColor": "#c8f4ff",
                        "border": "1px solid darkblue",
                    },
                ],
                export_format='xlsx',
                export_headers='ids',
                id='failureTable'
            ),
        )
    )


def getFailureAnalysisDirectionColumn(laminateObject):
    directionColumn = []
    for layerNo in np.arange(0, laminateObject.getNumberOfLaminateLayers()):
        for position in np.arange(0, 3):
```

```
            directionColumn.append(1)
            directionColumn.append(2)
            directionColumn.append(12)
    return directionColumn


def getFailureAnalysisLaminaeColumn(laminateObject):
    laminaeColumn = []
    for layerNo in np.arange(0, laminateObject.getNumberOfLaminateLayers
        ↪ ()):
        for laminaNo in np.arange(0, 3):
            laminaeColumn.append(layerNo + 1)
            laminaeColumn.append(layerNo + 1)
            laminaeColumn.append(layerNo + 1)
    return laminaeColumn


def getFailureAnalysisMaxStrainColumn(laminateObject):
    strain12Values = getStrain12Values(laminateObject)
    maxStrainColumn = []
    for layerNo in np.arange(0, laminateObject.getNumberOfLaminateLayers
        ↪ ()):
        if globalVars.isNotFailed(layerNo):
            currentLaminae = laminateObject.getLaminateLayer(layerNo)
            keps1t = currentLaminae.getProperties()["keps1t"]
            keps2t = currentLaminae.getProperties()["keps2t"]
            keps1c = currentLaminae.getProperties()["keps1c"]
            keps2c = currentLaminae.getProperties()["keps2c"]
            keps12s = currentLaminae.getProperties()["keps12s"]
            for position in np.arange(0, 3):
                strain1 = strain12Values[0][layerNo * 3 + position]
                strain2 = strain12Values[1][layerNo * 3 + position]
                strain12 = strain12Values[2][layerNo * 3 + position]
                if strain1 < 0:
                    keps1 = keps1c
                else:
                    keps1 = keps1t
                if strain2 < 0:
                    keps2 = keps2c
                else:
                    keps2 = keps2t
                maxStrainColumn.append(keps1 / strain1 if strain1 else
                    ↪ 0.0)
                maxStrainColumn.append(keps2 / strain2 if strain2 else
                    ↪ 0.0)
                maxStrainColumn.append(keps12s / abs(strain12) if
                    ↪ strain12 else 0.0)
        else:
            for position in np.arange(0, 3):
                maxStrainColumn.append(0.0)
                maxStrainColumn.append(0.0)
                maxStrainColumn.append(0.0)

    return maxStrainColumn


```

```
def getFailureAnalysisMaxStressColumn(laminateObject):
    stress12Values = getStress12Values(laminateObject)
    maxStressColumn = []
    for layerNo in np.arange(0, laminateObject.getNumberOfLaminateLayers
        ↪ ()):
        if globalVars.isNotFailed(layerNo):
            currentLaminae = laminateObject.getLaminateLayer(layerNo)
            k1t = currentLaminae.getProperties()["k1t"]
            k2t = currentLaminae.getProperties()["k2t"]
            k1c = currentLaminae.getProperties()["k1c"]
            k2c = currentLaminae.getProperties()["k2c"]
            k12s = currentLaminae.getProperties()["k12s"]
            for position in np.arange(0, 3):
                stress1 = stress12Values[0][layerNo * 3 + position]
                stress2 = stress12Values[1][layerNo * 3 + position]
                stress12 = stress12Values[2][layerNo * 3 + position]
                if stress1 < 0:
                    k1 = k1c
                else:
                    k1 = k1t
                if stress2 < 0:
                    k2 = k2c
                else:
                    k2 = k2t
                maxStressColumn.append(k1 / stress1 if stress1 else 0.0)
                maxStressColumn.append(k2 / stress2 if stress2 else 0.0)
                maxStressColumn.append(k12s / abs(stress12) if stress12
                    ↪ else 0.0)
        else:
            for position in np.arange(0, 3):
                maxStressColumn.append(0.0)
                maxStressColumn.append(0.0)
                maxStressColumn.append(0.0)
    return maxStressColumn


def getFailureAnalysisPositionColumn(laminateObject):
    positionColumn = []
    positionLabelArray = ["Bottom", "Middle", "Top"]
    for layerNo in np.arange(0, laminateObject.getNumberOfLaminateLayers
        ↪ ()):
        for laminaNo in np.arange(0, 3):
            positionColumn.append(positionLabelArray[laminaNo])
            positionColumn.append(positionLabelArray[laminaNo])
            positionColumn.append(positionLabelArray[laminaNo])
    return positionColumn


def getFailureAnalysisTsaiHillColumn(laminateObject):
    stress12Values = getStress12Values(laminateObject)
    tsaiHillColumn = []
    for currentLaminae in laminateObject:
        layerNo = currentLaminae.getLayerNo()
        if globalVars.isNotFailed(layerNo):
            currentLaminae = laminateObject.getLaminateLayer(layerNo)
            for position in np.arange(0, 3):
```

```
                stress1 = stress12Values[0][layerNo * 3 + position]
                stress2 = stress12Values[1][layerNo * 3 + position]
                stress12 = stress12Values[2][layerNo * 3 + position]
                tsaiHillColumn.append(0.0)
                tsaiHillColumn.append(getRTsaiHill(currentLaminae,
                    ↪ stress1, stress2, stress12))
                tsaiHillColumn.append(0.0)
        else:
            for position in np.arange(0, 3):
                tsaiHillColumn.append(0.0)
                tsaiHillColumn.append(0.0)
                tsaiHillColumn.append(0.0)
    return tsaiHillColumn


def getFailureAnalysisTsaiWuColumn(laminateObject):
    stress12Values = getStress12Values(laminateObject)
    tsaiWuColumn = []
    for layerNo in np.arange(0, laminateObject.getNumberOfLaminateLayers
        ↪ ()):
        if globalVars.isNotFailed(layerNo):
            currentLaminae = laminateObject.getLaminateLayer(layerNo)
            for position in np.arange(0, 3):
                stress1 = stress12Values[0][layerNo * 3 + position]
                stress2 = stress12Values[1][layerNo * 3 + position]
                stress12 = stress12Values[2][layerNo * 3 + position]
                tsaiWuColumn.append(0.0)
                tsaiWuColumn.append(getRTsaiWu(currentLaminae, stress1,
                    ↪ stress2, stress12))
                tsaiWuColumn.append(0.0)
        else:
            for position in np.arange(0, 3):
                tsaiWuColumn.append(0.0)
                tsaiWuColumn.append(0.0)
                tsaiWuColumn.append(0.0)
    return tsaiWuColumn


def getFailureEnvelopesContainer(laminateObject):
    firstLaminae = copy.deepcopy(laminateObject.getLaminateLayer(0))
    alphaValues = np.arange(1, 91, 1).tolist()

    sigmaXTension = 1.0
    stressXYTension = np.array([[sigmaXTension, 0.0, 0.0]])

    sigmaXCompression = -1.0
    stressXYCompression = np.array([[sigmaXCompression, 0.0, 0.0]])

    maximumStressTensionValues = []
    maximumStressCompressionValues = []
    maximumStrainTensionValues = []
    maximumStrainCompressionValues = []
    tsaiHillTensionValues = []
    tsaiHillCompressionValues = []
    tsaiWuValues = []

```

```
    failureModeMaximumStressCriterionTension = []
    failureModeMaximumStressCriterionCompression = []
    failureModeMaximumStrainCriterionTension = []
    failureModeMaximumStrainCriterionCompression = []

    for alpha in alphaValues:
        firstLaminae.setProperties("alpha", alpha)
        stressTransformationMatrix = firstLaminae.getStressTransformationMatrix()

        stress12Tension = stressTransformationMatrix.dot(stressXYTension.transpose()).transpose()
        stress12Compression = stressTransformationMatrix.dot(stressXYCompression.transpose()).transpose()

        complianceMatrix = firstLaminae.getComplianceMatrix()
        strain12Tension = complianceMatrix.dot(stress12Tension.transpose()).transpose()
        strain12Compression = complianceMatrix.dot(stress12Compression.transpose()).transpose()

        [rTension, failureMode] = getFailureEnvelopesRMaximumStressCriteria(firstLaminae, stress12Tension)
        failureModeMaximumStressCriterionTension.append(failureMode)
        maximumStressTensionValues.append(rTension)
        [rCompression, failureMode] = getFailureEnvelopesRMaximumStressCriteria(firstLaminae, stress12Compression)
        failureModeMaximumStressCriterionCompression.append(failureMode)
        maximumStressCompressionValues.append(rCompression)

        [rTension, failureMode] = getFailureEnvelopesRMaximumStrainCriteria(firstLaminae, strain12Tension)
        failureModeMaximumStrainCriterionTension.append(failureMode)
        maximumStrainTensionValues.append(rTension)
        [rCompression, failureMode] = getFailureEnvelopesRMaximumStrainCriteria(firstLaminae, strain12Compression)
        failureModeMaximumStrainCriterionCompression.append(failureMode)
        maximumStrainCompressionValues.append(rCompression)

        tsaiHillTensionValues.append(
            getRTsaiHill(firstLaminae, stress12Tension[0][0], stress12Tension[0][1], stress12Tension[0][2]))
        tsaiHillCompressionValues.append(
            getRTsaiHill(firstLaminae, stress12Compression[0][0], stress12Compression[0][1], stress12Compression[0][2]))
        tsaiWuValues.append(
            getRTsaiWu(firstLaminae, stress12Tension[0][0], stress12Tension[0][1], stress12Tension[0][2]))

    maximumStressPlots = {
        "tension": {
```

```
            "xValues": alphaValues,
            "yValues": maximumStressTensionValues,
            "title": r"$\text{tension} \ \sigma_{x} > 0$",
            "lineColor": "black",
            "lineWidth": 1,
            "markerSize": 6
        },
        "compression": {
            "xValues": alphaValues,
            "yValues": maximumStressCompressionValues,
            "title": r"$\text{compression} \ \sigma_{x} < 0$",
            "lineColor": "darkred",
            "lineWidth": 1,
            "markerSize": 6
        },
        "plotXMax": 90,
        "plotYMax": max(max(maximumStressTensionValues), max(maximumStressCompressionValues)) * 1.1,
        "xType": r"$\text{Off-axis angle} \ \alpha$",
        "yType": r"$\text{Normal stress} \ |{\sigma_{x}}| \text{ in MPa}$",
    }

    maximumStrainPlots = {
        "tension": {
            "xValues": alphaValues,
            "yValues": maximumStrainTensionValues,
            "title": r"$\text{tension} \ \sigma_{x} > 0$",
            "lineColor": "black",
            "lineWidth": 1,
            "markerSize": 6
        },
        "compression": {
            "xValues": alphaValues,
            "yValues": maximumStrainCompressionValues,
            "title": r"$\text{compression} \ \sigma_{x} < 0$",
            "lineColor": "darkred",
            "lineWidth": 1,
            "markerSize": 6
        },
        "plotXMax": 90,
        "plotYMax": max(max(maximumStrainTensionValues), max(maximumStrainCompressionValues)) * 1.1,
        "xType": r"$\text{Off-axis angle} \ \alpha$",
        "yType": r"$\text{Normal stress} \ |{\sigma_{x}}| \text{ in MPa}$",
    }

    tsaiHillPlots = {
        "tension": {
            "xValues": alphaValues,
            "yValues": tsaiHillTensionValues,
            "title": r"$\text{tension} \ \sigma_{x} > 0$",
            "lineColor": "black",
            "lineWidth": 1,
            "markerSize": 6
```

```
        },
        "compression": {
            "xValues": alphaValues,
            "yValues": tsaiHillCompressionValues,
            "title": r"$\text{compression} \ \sigma_{x} < 0$",
            "lineColor": "darkred",
            "lineWidth": 1,
            "markerSize": 6
        },
        "plotXMax": 90,
        "plotYMax": max(max(tsaiHillTensionValues), max(tsaiHillCompressionValues)) * 1.1,
        "xType": r"$\text{Off-axis angle} \ \alpha$",
        "yType": r"$\text{Normal stress} \ |{\sigma_{x}}| \text{ in MPa}$",
    }

    tsaiWuPlot = {
        "tension": {
            "xValues": alphaValues,
            "yValues": tsaiWuValues,
            "title": r"$\text{Tsai-Wu}$",
            "lineColor": "black",
            "lineWidth": 1,
            "markerSize": 6
        },
        "plotXMax": 90,
        "plotYMax": max(tsaiWuValues) * 1.1,
        "xType": r"$\text{Off-axis angle} \ \alpha$",
        "yType": r"$\text{Normal stress} \ |{\sigma_{x}}| \text{ in MPa}$",
    }

    maximumStressCriterionDataTableObject = getFailureModeTable(
        alphaValues,
        failureModeMaximumStressCriterionTension,
        failureModeMaximumStressCriterionCompression
    )

    maximumStrainCriterionDataTableObject = getFailureModeTable(
        alphaValues,
        failureModeMaximumStrainCriterionTension,
        failureModeMaximumStrainCriterionCompression
    )

    containerRows = []

    containerRows.append(
        dcc.Markdown('''#### Failure envelope for a unidirectional prepreg''',
                     mathjax=True)
    )
    containerRows.append(html.Br())

    containerRows.append(
        dcc.Markdown('''##### Maximum stress criterion''',
```

```
                mathjax=True)
    )

    containerRows.append(html.Br())
    containerRows.append(
        dcc.Markdown('''###### Failure Modes''',
                mathjax=True)
    )

    containerRows.append(html.Br())
    containerRows.append(maximumStressCriterionDataTableObject)

    containerRows.append(html.Br())

    containerRows.append(dmc.Center(
        getCustomizedGraph(makeLinePlot(maximumStressPlots), {"fileName"
            : "maximumStressPlots"})
    ))
    maximumStressPlots.update({"logY": True,
                               "yType": r"$\text{Normal stress} \ |\text{lg} \, {\sigma_{x}}| \text{ in MPa}$",
                               "plotYMax": np.log10(
                                   max(max(maximumStressTensionValues), max(maximumStressCompressionValues)) * 1.1),
                               "plotYMin": np.log10(
                                   min(min(maximumStressTensionValues), min(maximumStressCompressionValues)) * 0.9),
                               }
                              )
    containerRows.append(dmc.Center(
        getCustomizedGraph(makeLinePlot(maximumStressPlots), {"fileName"
            : "maximumStressLogPlots"})
    ))

    tableData = {
        "Angle": {
            "title": "Angle",
            "precision": 1
        },
        "Tension": {
            "title": "Tension",
            "precision": 3
        },
        "TensionFailureMode": {
            "title": "Tension Failure Mode"
        },
        "Compression": {
            "title": "Compression",
            "precision": 3
        },
        "CompressionFailureMode": {
            "title": "Compression Failure Mode"
        },
```

```
        "Tsai-Wu": {
            "title": "Tsai-Wu",
            "precision": 3
        }
    }

    diagramValues = getTableDataFrame([pd.DataFrame(alphaValues),
                    pd.DataFrame(maximumStressTensionValues),
                    pd.DataFrame(failureModeMaximumStressCriterionTension),
                    pd.DataFrame(maximumStressCompressionValues),
                    pd.DataFrame(failureModeMaximumStressCriterionCompression)],
                                    ['Angle',
                                     'Tension', 'Tension Failure Mode',
                                     'Compression', 'Compression Failure Mode'])

    containerRows.append(
        getAccordionDataTable(tableData, diagramValues)
    )
    containerRows.append(html.Hr())

    containerRows.append(
        dcc.Markdown('##### Maximum strain criterion')
    )

    containerRows.append(html.Br())
    containerRows.append(
        dcc.Markdown('###### Failure Modes')
    )

    containerRows.append(html.Br())
    containerRows.append(maximumStrainCriterionDataTableObject)

    containerRows.append(html.Br())

    containerRows.append(dmc.Center(
        getCustomizedGraph(makeLinePlot(maximumStrainPlots), {"fileName": "maximumStrainPlots"})
    ))
    maximumStrainPlots.update({"logY": True,
                               "yType": r"$\text{Normal stress} \ | \text{lg} \, {\sigma_{x}} | \text{ in MPa}$",
                               "plotYMax": np.log10(
                                   max(max(maximumStrainTensionValues),max(maximumStrainCompressionValues)) * 1.1),
                               "plotYMin": np.log10(
                                   min(min(maximumStrainTensionValues),min(maximumStrainCompressionValues)) * 0.9),
                               }
                              )
    containerRows.append(dmc.Center(
```

```
        getCustomizedGraph(makeLinePlot(maximumStrainPlots), {"fileName": "maximumStrainLogPlots"})
    ))

    diagramValues = getTableDataFrame([pd.DataFrame(alphaValues),
                    pd.DataFrame(maximumStrainTensionValues),
                    pd.DataFrame(failureModeMaximumStrainCriterionTension),
                    pd.DataFrame(maximumStrainCompressionValues),
                    pd.DataFrame(failureModeMaximumStrainCriterionCompression)],
                                      ['Angle',
                                       'Tension', 'Tension Failure Mode',
                                       'Compression', 'Compression Failure Mode'])
    containerRows.append(
        getAccordionDataTable(tableData, diagramValues)
    )
    containerRows.append(html.Hr())

    # Tsai Hill

    containerRows.append(
        dcc.Markdown('##### Tsai-Hill criterion')
    )

    containerRows.append(html.Br())

    containerRows.append(dmc.Center(
        getCustomizedGraph(makeLinePlot(tsaiHillPlots), {"fileName": "tsaiHillPlots"})
    ))

    tsaiHillPlots.update({"logY": True,
                          "yType": r"$\text{Normal stress} \ | \text{lg} \, {\sigma_{x}} | \text{ in MPa}$",
                          "plotYMax": np.log10(
                              max(max(tsaiHillTensionValues), max(tsaiHillCompressionValues)) * 1.1),
                          "plotYMin": np.log10(
                              min(min(tsaiHillTensionValues), min(tsaiHillCompressionValues)) * 0.9),
                          }
                         )
    containerRows.append(dmc.Center(
        getCustomizedGraph(makeLinePlot(tsaiHillPlots), {"fileName": "tsaiHillLogPlots"})
    ))

    diagramValues = getTableDataFrame([pd.DataFrame(alphaValues),
                                       pd.DataFrame(tsaiHillTensionValues),
                                       pd.DataFrame(tsaiHillCompressionValues)],
                                      ['Angle', 'Tension', 'Compression'])
    containerRows.append(
        getAccordionDataTable(tableData, diagramValues)
```

```python
    )
    containerRows.append(html.Hr())

    # Tsai Wu

    containerRows.append(
        dcc.Markdown('##### Tsai-Wu criterion')
    )

    containerRows.append(html.Br())

    containerRows.append(dmc.Center(
        getCustomizedGraph(makeLinePlot(tsaiWuPlot), {"fileName": "tsaiWuPlots"})
    ))

    tsaiWuPlot.update({"logY": True,
                       "yType": r"$\text{Normal stress} \ | \text{lg} \, {\sigma_{x}} | \text{ in MPa}$",
                       "plotYMax": np.log10(max(tsaiWuValues) * 1.1),
                       "plotYMin": np.log10(min(tsaiWuValues) * 0.9),
                       }
                      )
    containerRows.append(dmc.Center(
        getCustomizedGraph(makeLinePlot(tsaiWuPlot), {"fileName": "tsaiWuLogPlots"})
    ))

    diagramValues = getTableDataFrame([pd.DataFrame(alphaValues),
                                       pd.DataFrame(tsaiWuValues)],
                                      ['Angle', 'Tsai-Wu'])
    containerRows.append(
        getAccordionDataTable(tableData, diagramValues)
    )
    containerRows.append(html.Hr())

    return dbc.Container(containerRows)


def getFailureEnvelopesRMaximumStrainCriteria(laminae, cVector):
    keps1t = laminae.getProperties()["keps1t"]
    keps2t = laminae.getProperties()["keps2t"]
    keps12s = laminae.getProperties()["keps12s"]
    keps1c = laminae.getProperties()["keps1c"]
    keps2c = laminae.getProperties()["keps2c"]
    [[c1, c2, c3]] = cVector

    # tension
    if c1 >= 0.0:
        r1 = keps1t / c1 if c1 else 0.0
    # compression
    else:
        r1 = keps1c / c1

    # tension
    if c2 >= 0.0:
```

```
        r2 = keps2t / c2 if c2 else 0.0
    # compression
    else:
        r2 = keps2c / c2

    r3 = abs(keps12s / c3 if c3 else 0.0)

    # print(c1, " ", r1, r2, r3)
    R = min(r1, r2, r3)

    failureMode = "not known"
    # tension
    if c1 >= 0.0:
        if round(c1 * R, 5) >= keps1t:
            failureMode = "Longitudinal tensile failure (k1t)"
    # compression
    else:
        if round(c1 * R, 5) <= keps1c:
            failureMode = "Longitudinal compression failure (k1c)"

    # tension
    if c2 >= 0.0:
        if round(c2 * R, 5) >= keps2t:
            failureMode = "Transverse tensile failure (k2t)"
    # compression
    else:
        if round(c2 * R, 5) <= keps2c:
            failureMode = "Transverse compression failure (k2c)"

    if round(abs(c3 * R), 5) >= keps12s:
        failureMode = "Shear failure (k12s)"

    return [min(r1, r2, r3), failureMode]


def getFailureEnvelopesRMaximumStressCriteria(laminae, cVector):
    k1t = laminae.getProperties()["k1t"]
    k2t = laminae.getProperties()["k2t"]
    k12s = laminae.getProperties()["k12s"]
    k1c = laminae.getProperties()["k1c"]
    k2c = laminae.getProperties()["k2c"]
    [[c1, c2, c3]] = cVector

    # tension
    if c1 >= 0.0:
        r1 = k1t / c1 if c1 else 0.0
    # compression
    else:
        r1 = k1c / c1

    # tension
    if c2 >= 0.0:
        r2 = k2t / c2 if c2 else 0.0
    # compression
    else:
        r2 = k2c / c2
```

```

    r3 = abs(k12s / c3 if c3 else 0.0)
    # print(c1, " ", r1, r2, r3)

    R = min(r1, r2, r3)

    # tension
    if c1 >= 0.0:
        if round(c1 * R, 5) >= k1t:
            failureMode = "Longitudinal tensile failure (k1t)"
    # compression
    else:
        if round(c1 * R, 5) <= k1c:
            failureMode = "Longitudinal compression failure (k1c)"

    # tension
    if c2 >= 0.0:
        if round(c2 * R, 5) >= k2t:
            failureMode = "Transverse tensile failure (k2t)"
    # compression
    else:
        if round(c2 * R, 5) <= k2c:
            failureMode = "Transverse compression failure (k2c)"

    if round(abs(c3 * R), 5) >= k12s:
        failureMode = "Shear failure (k12s)"

    return [min(r1, r2, r3), failureMode]


def getFailureModeTable(
        alphaValues,
        failureModeTension,
        failureModeCompression):
    firstColumn = []
    secondColumn = []
    # Tension
    firstColumn.append("Tension (sigma_x > 0)")
    secondColumn.append("")
    index = 1
    startFailureMode = copy.copy(failureModeTension[0])
    startAngle = 0
    while index < len(alphaValues) - 1:
        index += 1
        if index == len(alphaValues) - 1:
            firstColumn.append(str(startAngle) + "  0  
                ↪ " + str(alphaValues[index]))
            secondColumn.append(startFailureMode)
        elif failureModeTension[index] != startFailureMode:
            firstColumn.append(str(startAngle) + "  0  
                ↪ " + str(alphaValues[index - 1]))
            secondColumn.append(startFailureMode)
            startFailureMode = copy.copy(failureModeTension[index])
            startAngle = copy.copy(alphaValues[index])
    firstColumn.append("")
    secondColumn.append("")
```

```python
    # Compression
    firstColumn.append("Compression (sigma_x < 0)")
    secondColumn.append("")
    index = 1
    startFailureMode = copy.copy(failureModeCompression[0])
    startAngle = 0
    while index < len(alphaValues) - 1:
        index += 1
        if index == len(alphaValues) - 1:
            firstColumn.append(str(startAngle) + "  0  " + str(alphaValues[index]))
            secondColumn.append(startFailureMode)
        elif failureModeCompression[index] != startFailureMode:
            firstColumn.append(str(startAngle) + "  0  " + str(alphaValues[index - 1]))
            secondColumn.append(startFailureMode)
            startFailureMode = copy.copy(failureModeCompression[index])
            startAngle = copy.copy(alphaValues[index])

    tableDataFrame = getTableDataFrame(
        [pd.DataFrame(firstColumn),
         pd.DataFrame(secondColumn)],
        ['Angle range', 'Failure mode'])

    tableData = {
        'firstColumn': {
            'title': 'Angle range',
            'type': 'text'
        },
        'secondColumn': {
            'title': 'Failure mode',
            'type': 'text'
        },
    }
    return getDataTableObject(tableData, tableDataFrame)


def getLaminateSketch(data, userSelectedFailedPlies):
    for i in np.arange(0, len(data)):
        data[i]["thickness"] = data[i]["zK"] - data[i]["zKminusOne"]
        data[i]['title'] = 'Layer {0: <2}'.format(str(data[i]['Layer'])) + ' ({:+02.1f})'.format((data[i]['alpha']))

    for i in userSelectedFailedPlies:
        data[i]["alpha"] = 10000

    if not userSelectedFailedPlies:
        colorBar = px.colors.sequential.gray
    else:
        colorBar = px.colors.sequential.Jet

    laminateSketch = px.bar(
        data, x="keps12s", y="thickness",
        color="alpha",
        color_continuous_scale=colorBar,
        template="simple_white",
```

```
        text="title",
        hover_data=['Layer', 'E1', 'E2', 'G12', 'v12', 'v21', 'k1t', '
            ↪ k1c', 'k2t', 'k2c', 'k12s', 'keps1t',
                'keps1c', 'keps2t', 'keps2c', 'keps12s', 'zKminusOne'
                    ↪ , 'zK', 'alpha'],
        height=600,
        width=700,
    )
    layout = {
        'xaxis': {
            'showgrid': False, # thin lines in the background
            'zeroline': False, # thick line at x=0
            'visible': False, # numbers below
        }, # the same for yaxis
        'yaxis': {
            'showgrid': False, # thin lines in the background
            'zeroline': False, # thick line at x=0
            'visible': False, # numbers below,
        }, # the same for yaxis
        'showlegend': False,
        'barmode': 'stack',

    }
    laminateSketch.update_traces(
        {
            'insidetextanchor': 'middle',
            'textposition': 'inside',
            'marker_line_color': 'rgb(0,0,0)',
            'marker_line_width': 1.5,
            'opacity': 1
        }
    )
    laminateSketch.update_layout(layout)
    laminateSketch.update_coloraxes(showscale=False)
    return laminateSketch


def getLaminateThicknessTable(dataFrame):
    return [
        html.H3("Laminate␣Thickness"),
        dash_table.DataTable(
            data=dataFrame.to_dict('records'),
            columns=[
                dict(id='Thickness', name="Thickness", type='numeric',
                    ↪ format=Format(precision=2, scheme=Scheme.fixed)),
            ],
            id={'type': 'table', 'index': 'thicknessMatrix'},
            style_header={
                'textAlign': 'center'
            },
            style_cell={
                'textAlign': 'center',
                'height': '32px'
            },
            style_data={
                'color': 'black',
```

```
                'backgroundColor': 'rgb(220, 220, 220)'
            },
            style_data_conditional=[
                {
                    "if": {"state": "selected"}, # 'active' | 'selected'
                    "backgroundColor": "#c8f4ff",
                    "border": "1px solid darkblue",
                },
            ],
            editable=True,
            page_action='native',
            page_current=0
        )
    ]


def getLaminateThicknessTableWithTooltip(dataFrame):
    return dmc.Tooltip(
        multiline=True,
        width="auto",
        withArrow=True,
        transition="fade",
        position="bottom",
        # placement="center",
        color="gray",
        transitionDuration=500,
        label=[
            html.P(
                'The "Laminate Thickness" value only affects the results for the ply-by-ply failure analysis.')
        ],
        children=getLaminateThicknessTable(dataFrame)
    )


def getLaminateThicknessTableWithTooltipColumn(dataFrame):
    return dbc.Col(
        getLaminateThicknessTableWithTooltip(dataFrame),
        width=6
    )


def getLayerColumn(laminateObject):
    layerColumn = []
    for layerNo in np.arange(0, laminateObject.getNumberOfLaminateLayers()):
        layerColumn.append(layerNo + 1)
        layerColumn.append(layerNo + 1)
        layerColumn.append(layerNo + 1)
    return layerColumn


def getLoadExistingModelsModule(files):
    return html.Div(
        id='dropdownParent',
        children=dcc.Dropdown(
```

```
            files,
            id="dropdown-menu",
            placeholder="Existing models...",
        ),
        style={
            'width': 180
        }
    )


def getLoadExistingModelsModuleWithTooltip(files):
    return dmc.Tooltip(
        multiline=True,
        width="auto",
        withArrow=True,
        transition="fade",
        position="bottom",
        # placement="center",
        color="gray",
        transitionDuration=500,
        label=[
            html.P(
                'The "Load existing models..." option checks the main folder for any existing Excel files.'
            ),
        ],
        children=getLoadExistingModelsModule(files)
    )


def getLoadExistingModelsModuleWithTooltipColumn(files):
    return dmc.Col(
        getLoadExistingModelsModuleWithTooltip(files),
        span=2
    )


def getLoadTable(dataFrame):
    return [
        html.H3("Load Matrix"),
        dash_table.DataTable(
            data=dataFrame.to_dict('records'),
            columns=[
                dict(id='N_x', name="N_x", type='numeric', format=Format(precision=2, scheme=Scheme.fixed)),
                dict(id='N_y', name="N_y", type='numeric', format=Format(precision=2, scheme=Scheme.fixed)),
                dict(id='N_xy', name="N_xy", type='numeric', format=Format(precision=2, scheme=Scheme.fixed)),
                dict(id='M_x', name="M_x", type='numeric', format=Format(precision=2, scheme=Scheme.fixed)),
                dict(id='M_y', name="M_y", type='numeric', format=Format(precision=2, scheme=Scheme.fixed)),
                dict(id='M_xy', name="M_xy", type='numeric', format=Format(precision=2, scheme=Scheme.fixed)),
            ],
```

```
                id={'type': 'table', 'index': 'myLoadMatrixTable'},
                style_header={
                    'textAlign': 'center'
                },
                style_cell={
                    'textAlign': 'center',
                    'height': '32px'
                },
                style_data={
                    'color': 'black',
                    'backgroundColor': 'rgb(220, 220, 220)'
                },
                style_data_conditional=[
                    {
                        "if": {"state": "selected"}, # 'active' | 'selected'
                        "backgroundColor": "#c8f4ff",
                        "border": "1px solid darkblue",
                    },
                ],
                editable=True,
                page_action='native',
                page_current=0
            )
        ]


def getLoadTableColumn(dataFrame):
    return dbc.Col(
        getLoadTable(dataFrame),
        width=5
    )


def getLoadingOverlay(dataFrame, matrixDataFrame, scalarDataFrame):
    return html.Div(
        style={
            "width": '100%'
        },
        children=dmc.LoadingOverlay(
            [
                html.H3("Import/Export Model Data"),
                dmc.Center(getUploadButtons()),
                dmc.Center(getDownloadButtons()),
                html.Br(),
                html.H3("Model Properties"),
                getDataTableWithTooltip(dataFrame),
                html.Br(),
                html.Br(),
                getDataTableManipulationRow(dataFrame),
                html.Br(),
                dbc.Row([
                    getLoadTableColumn(matrixDataFrame),
                    getLaminateThicknessTableWithTooltipColumn(
                        ↪ scalarDataFrame)
                ]),
                html.Br()
```

```
            ],
            id="loading-output",
            loaderProps={"variant": "dots", "color": "blue", "size": "xl"},
            radius=15)
    )


def getPlyByPlyAnalysis(laminateObject, userSelectedFailedPlies):
    maxStressGraphData = [0.0]
    maxStrainGraphData = [0.0]
    containerRows = []

    step = 0
    if userSelectedFailedPlies:
        containerRows.append(dbc.Row(html.Div(html.H2(f"Step {step}"))))
        globalVars.failedPlies = userSelectedFailedPlies
        containerRows.append(dbc.Row(html.Div(html.P("The user has chosen the following layers as pre-failed: " + str(
            [i + 1 for i in sorted(userSelectedFailedPlies)])))))
        stressPlots = getStressPlots(laminateObject)
        strainPlots = getStrainPlots(laminateObject)
        containerRowObjects = [
                                 dbc.Col(
                                     getCustomizedGraph(
                                         stressPlots["stressesXY"]["stressX"],
                                         {"fileName": "StressXPlot_Step_" + str(step)}
                                     ),
                                     width=6
                                 )
                             ] + [
                                 dbc.Col(
                                     getCustomizedGraph(
                                         strainPlots["strainsXY"]["strainX"],
                                         {"fileName": "StrainXPlot_Step_" + str(step)}
                                     ),
                                     width=6
                                 )
                             ]
        containerRows.append(dbc.Row(containerRowObjects))
    else:
        containerRows.append(dbc.Row(html.Div(html.H2(f"Step {step}"))))
        containerRows.append(dbc.Row(html.Div(html.P("This is the original configuration of the problem."))))
        globalVars.failedPlies = []
        stressPlots = getStressPlots(laminateObject)
        strainPlots = getStrainPlots(laminateObject)
        containerRowObjects = [
                                 dbc.Col(
                                     getCustomizedGraph(
```

```
                                        stressPlots["stressesXY"]["stressX"],
                                        {"fileName": "StressXPlot_Step_" + str(step)}
                                    ),
                                    width=6
                                )
                            ] + [
                                dbc.Col(
                                    getCustomizedGraph(
                                        strainPlots["strainsXY"]["strainX"],
                                        {"fileName": "StrainXPlot_Step_" + str(step)}
                                    ),
                                    width=6
                                )
                            ]
        containerRows.append(
            dbc.Row(containerRowObjects)
        )

    containerRows.append(
        dbc.Row(
            dbc.Col(
                html.Hr(style={'borderWidth': "0.3vh", "width": "100%", "backgroundColor": "#000000", "opacity": "1"}),
                width={'size': 10, 'offset': 1}
            )
        )
    )

    while len(globalVars.failedPlies) < laminateObject.getNumberOfLaminateLayers():

        step += 1
        containerRows.append(dbc.Row(html.Div(html.H2(f"Step {step}"))))
        tsaiWuColumn = getFailureAnalysisTsaiWuColumn(laminateObject)
        tsaiWuColumn = np.array(tsaiWuColumn)
        tsaiWuColumnNonZero = np.nonzero(tsaiWuColumn)
        tsaiWuMin = np.min(tsaiWuColumn[tsaiWuColumnNonZero])

        layerColumn = getFailureAnalysisLaminaeColumn(laminateObject)
        before = copy.deepcopy(globalVars.failureAnalysisFailedPlies)
        for row in np.arange(0, len(tsaiWuColumn)):
            if (tsaiWuMin - 1e-10) < tsaiWuColumn[row] < (tsaiWuMin + 1e-10):
                globalVars.failureAnalysisFailedPlies.append(layerColumn[row] - 1)

        maxStressX = tsaiWuMin * laminateObject.getLoadMatrix().tolist()[0][0] / laminateObject.getThickness()
        maxStressGraphData.append(float(maxStressX))
        maxStrainX = tsaiWuMin * laminateObject.getGeneralizedStrains()[0][0]
```

```
        maxStrainGraphData.append(float(maxStrainX) * 100)

        globalVars.failureAnalysisFailedPlies = list(set(globalVars.failureAnalysisFailedPlies))
        globalVars.failedPlies = userSelectedFailedPlies + globalVars.failureAnalysisFailedPlies

        containerRows.append(
            dbc.Row(
                html.Div(
                    html.P(
                        f"With the strength ratio of {tsaiWuMin:.3f}, failure occurred for layer number(s) \
                        {[i + 1 for i in sorted(set(globalVars.failureAnalysisFailedPlies) - set(before))]}."
                    )
                )
            )
        )

        stressPlots = getStressPlots(laminateObject)
        strainPlots = getStrainPlots(laminateObject)
        containerRowObjects = [
                                    dbc.Col(
                                        getCustomizedGraph(
                                            stressPlots["stressesXY"]["stressX"],
                                            {"fileName": "StressXPlot_Step_" + str(step)}
                                        ),
                                        width=6
                                    )
                                ] + [
                                    dbc.Col(
                                        getCustomizedGraph(
                                            strainPlots["strainsXY"]["strainX"],
                                            {"fileName": "StrainXPlot_Step_" + str(step)}
                                        ),
                                        width=6
                                    )
                                ]
        containerRows.append(dbc.Row(containerRowObjects))

        containerRows.append(
            dbc.Row(
                html.Div(
                    html.P(
                        "At the end of this step, the following layers have already been failed: " + str(
                    [i + 1 for i in sorted(set(globalVars.failedPlies))])
                    )
                )
            )
```

```
        )

        containerRows.append(dbc.Row(dbc.Col(
            html.Hr(style={'borderWidth': "0.3vh", "width": "100%", "
                ↪ backgroundColor": "#000000", "opacity": "1"}),
            width={'size': 10, 'offset': 1}), ))

    stressStrainPlot = {
        "stressStrainPlot": {
            "xValues": maxStrainGraphData,
            "yValues": maxStressGraphData,
            "title": r"$\text{tension}␣\␣\sigma_{x}␣>␣0$",
            "lineColor": "black",
            "lineWidth": 1,
            "markerSize": 6
        },
        "plotXMax": max(maxStrainGraphData) * 1.1,
        "plotYMax": max(maxStressGraphData) * 1.1,
        "xType": r"$\varepsilon_{x}␣\text{(%)}$",
        "yType": r"$N_x^n/t␣\,␣\text{(MPa)}$",
    }
    # graph = makeStressStrainPlot(graphData)
    graph = makeLinePlot(stressStrainPlot)
    containerRows.append(dbc.Row(html.Div(html.H2("Stress-Strain␣
        ↪ Diagram"))))
    containerRows.append(dbc.Row(dbc.Col(
        getCustomizedGraph(graph, {"fileName": "stressStrainPlot"}) ,
        width=6)))

    tableData = {
        "Strain": {
            "title": "Strain",
            "precision": 3
        },
        "Stress": {
            "title": "Stress",
            "precision": 3
        }
    }

    graphTable = getTableDataFrame(
        [
            pd.DataFrame(maxStrainGraphData),
            pd.DataFrame(maxStressGraphData),
        ],
        ['Strain', 'Stress']
    )

    containerRows.append(
        getAccordionDataTable(tableData, graphTable)
    )

    container = dbc.Container(containerRows)

    return container

```

```

def getPositionArray(laminae):
    zK = laminae.getZK()
    zKminusOne = laminae.getZKminusOne()
    topZ = zK - 0.00000001 # *abs(zK)
    bottomZ = zKminusOne + 0.00000001 # *abs(zKminusOne)
    middleZ = (zK + zKminusOne) / 2
    return [bottomZ, middleZ, topZ]


def getPositionColumn(laminateObject):
    positionColumn = []
    for layerNo in np.arange(0, laminateObject.getNumberOfLaminateLayers
        ↪ ()):
        positionColumn.append("Bottom")
        positionColumn.append("Center")
        positionColumn.append("Top")
    return positionColumn


def getRTsaiHill(laminae, stress1, stress2, stress12):
    k1t = laminae.getProperties()["k1t"]
    k1c = laminae.getProperties()["k1c"]
    k2t = laminae.getProperties()["k2t"]
    k2c = laminae.getProperties()["k2c"]
    k12s = laminae.getProperties()["k12s"]

    # Tsai-Hill
    if stress1 >= 0:
        k1TsaiHill = k1t
    else:
        k1TsaiHill = k1c

    if stress2 >= 0:
        k2TsaiHill = k2t
    else:
        k2TsaiHill = k2c

    R = SympySymbol('R', real=True)

    return float(solve(
        (R * stress1) ** 2 / k1TsaiHill ** 2
        -
        (R * stress1) * (R * stress2) / k1TsaiHill ** 2
        +
        (R * stress2) ** 2 / k2TsaiHill ** 2
        +
        (R * stress12) ** 2 / k12s ** 2
        - 1,
        R
    )[-1])


def getRTsaiWu(laminae, stress1, stress2, stress12):
    k1t = laminae.getProperties()["k1t"]
    k1c = laminae.getProperties()["k1c"]
```

```
    k2t = laminae.getProperties()["k2t"]
    k2c = laminae.getProperties()["k2c"]
    k12s = laminae.getProperties()["k12s"]

    # Tsai-Wu
    a1 = (1 / k1t) + (1 / k1c)
    a2 = (1 / k2t) + (1 / k2c)
    a11 = -1 / (k1t * k1c)
    a22 = -1 / (k2t * k2c)
    a12 = 1 / k12s ** 2
    a1comma2 = -(1 / 2) * np.sqrt((1 / (k1t * k1c)) * (1 / (k2t * k2c)))

    R = SympySymbol('R', real=True)

    return float(solve(
        a1 * (R * stress1) + a2 * (R * stress2) + a11 * (R * stress1) ** 2
        + a22 * (R * stress2) ** 2 + a12 * (R * stress12) ** 2
        + 2 * a1comma2 * (R * stress1) * (R * stress2) - 1,
        R
    )[-1])


def getStrain12DataFrame(laminateObject):
    strain12Values = getStrain12Values(laminateObject)
    dataFrame = pd.DataFrame(strain12Values).transpose()
    positionColumn = getPositionColumn(laminateObject)
    layerColumn = getLayerColumn(laminateObject)
    dataFrame.insert(0, "Position", positionColumn, True)
    dataFrame.insert(0, "Layer no.", layerColumn, True)
    dataFrame.columns = ['Layer no.', 'Position', 'eps_1', 'eps_2', 'eps_12']
    return dataFrame.to_dict('records')


def getStrain12Values(laminateObject):
    strain12Values = [[], [], []]
    for laminaeNumber in np.arange(0, laminateObject.getNumberOfLaminateLayers()):
        currentLaminae = laminateObject.getLaminateLayer(laminaeNumber)
        positionArray = getPositionArray(currentLaminae)
        for yValue in positionArray:
            strainAtY = laminateObject.getStrainsIn12(yValue).tolist()
            for index in np.arange(0, 3):
                strain12Values[index].append(strainAtY[index][0])
    return strain12Values


def getStrainAnalysisContainer(laminateObject):
    strain12Values = getStrain12Values(laminateObject)
    strainXYValues = getStrainXYValues(laminateObject)
    positionColumn = getPositionColumn(laminateObject)
    layerColumn = getLayerColumn(laminateObject)
    strainTableValues = getTableDataFrame([pd.DataFrame(layerColumn),
                                           pd.DataFrame(positionColumn),
```

```
                                        pd.DataFrame(strain12Values).transpose(),
                                        pd.DataFrame(strainXYValues).transpose()],
                                       ['Layer no.', 'Position',
                                        'eps_1', 'eps_2', 'gamma_12',
                                        'eps_x', 'eps_y', 'gamma_xy'])

    strainPlotData = {
        'stress12Values': strain12Values,
        'stressXYValues': strainXYValues,
    }
    plotData = getStrainPlots(laminateObject, strainPlotData)
    return dbc.Container(
        [
            dbc.Row(
                [
                    dbc.Col(
                        getCustomizedGraph(plotData["strains12"]["strain1"], {"fileName": "strain1Plot"}),
                        width=6
                    ),
                    dbc.Col(
                        getCustomizedGraph(plotData["strainsXY"]["strainX"], {"fileName": "strainXPlot"}),
                        width=6
                    ),
                ]
            ),

            dbc.Row(
                [
                    dbc.Col(
                        getCustomizedGraph(plotData["strains12"]["strain2"], {"fileName": "strain2Plot"}),
                        width=6,
                    ),
                    dbc.Col(
                        getCustomizedGraph(plotData["strainsXY"]["strainY"], {"fileName": "strainYPlot"}),
                        width=6
                    ),
                ]
            ),
            dbc.Row(
                [
                    dbc.Col(
                        getCustomizedGraph(plotData["strains12"]["strain12"], {"fileName": "strain12Plot"}),
                        width=6
                    ),
                    dbc.Col(
                        getCustomizedGraph(plotData["strainsXY"]["strainXY"], {"fileName": "strainXYPlot"}),
                        width=6
                    ),
```

```
                ]
            ),

            dbc.Row(
                dbc.Accordion(
                    [
                        dbc.AccordionItem(
                            dash_table.DataTable(
                                data=strainTableValues,
                                columns=[
                                    dict(id='Layer no.', name=['Layer no.'],
                                         type='numeric', format=Format()),
                                    dict(id='Position',
                                         name=['Position'],
                                         type='text',
                                         format=Format()),
                                    dict(id='eps_1',
                                         name=['eps_1'],
                                         type='numeric',
                                         format=Format(precision=6,
                                                       scheme=Scheme.fixed)),
                                    dict(id='eps_2',
                                         name=['eps_2'],
                                         type='numeric',
                                         format=Format(precision=6,
                                                       scheme=Scheme.fixed)),
                                    dict(id='gamma_12',
                                         name=['gamma_12'],
                                         type='numeric',
                                         format=Format(precision=6,
                                                       scheme=Scheme.fixed)),
                                    dict(id='eps_x',
                                         name=['eps_x'],
                                         type='numeric',
                                         format=Format(precision=6,
                                                       scheme=Scheme.fixed)),
                                    dict(id='eps_y',
                                         name=['eps_y'],
                                         type='numeric',
                                         format=Format(precision=6,
                                                       scheme=Scheme.fixed)),
                                    dict(id='gamma_xy',
                                         name=['gamma_xy'],
                                         type='numeric',
                                         format=Format(precision=6,
                                                       scheme=Scheme.fixed))
                                ],
                                style_header={
                                    'textAlign': 'center'
                                },
                                style_cell={
                                    'textAlign': 'center'
                                },
                                style_data={
                                    'color': 'black',
```

```
                                    'backgroundColor': 'white'
                                },
                                style_data_conditional=[
                                    {
                                        'if': {
                                            # 'row_index': 'odd'
                                            'filter_query': '{Position} eq "Bottom"',
                                        },
                                        'backgroundColor': 'rgb(220, 220, 220)',
                                    },
                                    {
                                        "if": {"state": "selected"}, # 'active' | 'selected'
                                        "backgroundColor": "#c8f4ff",
                                        "border": "1px solid darkblue",
                                    },

                                ],
                                export_format='xlsx',
                                export_headers='ids',
                                id='tbl'
                            ),
                            title="Strain Table"
                        )
                    ],

                    start_collapsed=True,
                )
            ),

        ])


def getStrainPlots(laminateObject, strainPlotData={}):
    if strainPlotData:
        strain12Values = strainPlotData["stress12Values"]
        strainXYValues = strainPlotData["stressXYValues"]
    else:
        strain12Values = getStrain12Values(laminateObject)
        strainXYValues = getStrainXYValues(laminateObject)
    yValues = getYValues(laminateObject)

    for index in np.arange(0, 3):
        strain12Values[index].insert(0, 0.0)
        strainXYValues[index].insert(0, 0.0)
        strain12Values[index].append(0.0)
        strainXYValues[index].append(0.0)

    yValues.insert(0, yValues[0])
    yValues.append(yValues[-1])

    strainData = {
        "strain121": {
            "xValues": strain12Values[0],
```

```
            "yValues": yValues,
            "type": r"$\varepsilon_{\textit{1}}␣\text{(-)}$"
        },
        "strain122": {
            "xValues": strain12Values[1],
            "yValues": yValues,
            "type": r"$\varepsilon_{\textit{2}}␣\text{(-)}$"
        },
        "strain123": {
            "xValues": strain12Values[2],
            "yValues": yValues,
            "type": r"$\gamma_{\textit{12}}␣\text{(-)}$"
        },
        "strainXY1": {
            "xValues": strainXYValues[0],
            "yValues": yValues,
            "type": r"$\varepsilon_{x}␣\text{(-)}$"
        },
        "strainXY2": {
            "xValues": strainXYValues[1],
            "yValues": yValues,
            "type": r"$\varepsilon_{y}␣\text{(-)}$"
        },
        "strainXY3": {
            "xValues": strainXYValues[2],
            "yValues": yValues,
            "type": r"$\gamma_{xy}␣\text{(-)}$"
        }
    }

    strains121 = makePlot(strainData["strain121"])
    strains122 = makePlot(strainData["strain122"])
    strains123 = makePlot(strainData["strain123"])
    strainsXY1 = makePlot(strainData["strainXY1"])
    strainsXY2 = makePlot(strainData["strainXY2"])
    strainsXY3 = makePlot(strainData["strainXY3"])

    return {"strains12": {"strain1": strains121, "strain2": strains122, "strain12": strains123},
            "strainsXY": {"strainX": strainsXY1, "strainY": strainsXY2, "strainXY": strainsXY3}}


def getStrainXYValues(laminateObject):
    strainXYValues = [[], [], []]
    for laminaeNumber in np.arange(0, laminateObject.getNumberOfLaminateLayers()):
        currentLaminae = laminateObject.getLaminateLayer(laminaeNumber)
        positionArray = getPositionArray(currentLaminae)
        for yValue in positionArray:
            strainAtY = laminateObject.getStrainsInXY(yValue).tolist()
            for index in np.arange(0, 3):
                strainXYValues[index].append(strainAtY[index][0])
    return strainXYValues
```

```
def getStrengthRatioDataFrame(laminateObject):
    strengthRatioValues = getStrengthRatioValues(laminateObject)
    strengthRatioDataFrame = pd.DataFrame(strengthRatioValues).transpose()

    strengthRatioDataFrame.columns = ['Layer no.', 'Position', 'Direction', 'Stress Max', 'Strain Max', 'Tsai-Hill',
                                      'Tsai-Wu']

    return strengthRatioDataFrame.to_dict('records')


def getStrengthRatioValues(laminateObject):
    laminaeColumn = getFailureAnalysisLaminaeColumn(laminateObject)
    positionColumn = getFailureAnalysisPositionColumn(laminateObject)
    directionColumn = getFailureAnalysisDirectionColumn(laminateObject)
    maxStressColumn = getFailureAnalysisMaxStressColumn(laminateObject)
    maxStrainColumn = getFailureAnalysisMaxStrainColumn(laminateObject)
    tsaiHillColumn = getFailureAnalysisTsaiHillColumn(laminateObject)
    tsaiWuColumn = getFailureAnalysisTsaiWuColumn(laminateObject)

    return [laminaeColumn, positionColumn, directionColumn, maxStressColumn, maxStrainColumn,
            [float(x) for x in tsaiHillColumn], [float(x) for x in tsaiWuColumn]]


def getStress12DataFrame(laminateObject):
    stress12Values = getStress12Values(laminateObject)
    dataFrame = pd.DataFrame(stress12Values).transpose()
    positionColumn = getPositionColumn(laminateObject)
    layerColumn = getLayerColumn(laminateObject)
    dataFrame.insert(0, "Position", positionColumn, True)
    dataFrame.insert(0, "Layer no.", layerColumn, True)
    dataFrame.columns = ['Layer no.', 'Position', 'sigma_1', 'sigma_2', 'sigma_12']
    return dataFrame.to_dict('records')


def getStress12Values(laminateObject):
    # ("check point generate graphs", globalVars.failedPlies, globalVars.failureAnalysisFailedPlies)
    stress12Values = [[], [], []]
    for laminaeNumber in np.arange(0, laminateObject.getNumberOfLaminateLayers()):
        currentLaminae = laminateObject.getLaminateLayer(laminaeNumber)
        positionArray = getPositionArray(currentLaminae)
        for yValue in positionArray:
            stressAtY = laminateObject.getStressesIn12(yValue).tolist()
            for index in np.arange(0, 3):
                stress12Values[index].append(stressAtY[index][0])
    return stress12Values


def getStressAnalysisContainer(laminateObject):
    stress12Values = getStress12Values(laminateObject)
```

```
    stressXYValues = getStressXYValues(laminateObject)
    positionColumn = getPositionColumn(laminateObject)
    layerColumn = getLayerColumn(laminateObject)
    stressTableValues = getTableDataFrame(
        [
            pd.DataFrame(layerColumn),
            pd.DataFrame(positionColumn),
            pd.DataFrame(stress12Values).transpose(),
            pd.DataFrame(stressXYValues).transpose()
        ],
        ['Layer no.', 'Position',
         'sigma_1', 'sigma_2', 'sigma_12',
         'sigma_x', 'sigma_y', 'sigma_xy'
         ]
    )

    stressPlotData = {
        'stress12Values': stress12Values,
        'stressXYValues': stressXYValues,
    }
    plotData = getStressPlots(laminateObject, stressPlotData)
    return dbc.Container(
        [
            dbc.Row(
                [
                    dbc.Col(
                        getCustomizedGraph(plotData["stresses12"]["stress1"], {"fileName": "stress1Plot"}),
                        width=6
                    ),
                    dbc.Col(
                        getCustomizedGraph(plotData["stressesXY"]["stressX"], {"fileName": "stressXPlot"}),
                        width=6
                    ),
                ]
            ),

            dbc.Row(
                [
                    dbc.Col(
                        getCustomizedGraph(plotData["stresses12"]["stress2"], {"fileName": "stress2Plot"}),
                        width=6,
                    ),
                    dbc.Col(
                        getCustomizedGraph(plotData["stressesXY"]["stressY"], {"fileName": "stressYPlot"}),
                        width=6
                    ),
                ]
            ),
            dbc.Row(
                [
                    dbc.Col(
```

```
                        getCustomizedGraph(plotData["stresses12"]["
                            ↪ stress12"], {"fileName": "stress12Plot"}),
                        width=6
                    ),
                    dbc.Col(
                        getCustomizedGraph(plotData["stressesXY"]["
                            ↪ stressXY"], {"fileName": "stressXYPlot"}),
                        width=6
                    ),
                ]
            ),

            dbc.Row(
                dbc.Accordion(
                    [
                        dbc.AccordionItem(
                            dash_table.DataTable(
                                data=stressTableValues,
                                columns=[
                                    dict(id='Layer␣no.', name=['Layer␣
                                        ↪ no.'],
                                        type='numeric', format=Format()),
                                    dict(id='Position',
                                        name=['Position'],
                                        type='text',
                                        format=Format()),
                                    dict(id='sigma_1',
                                        name=['sigma_1'],
                                        type='numeric',
                                        format=Format(precision=3,
                                                    scheme=Scheme.fixed)),
                                    dict(id='sigma_2',
                                        name=['sigma_2'],
                                        type='numeric',
                                        format=Format(precision=3,
                                                    scheme=Scheme.fixed)),
                                    dict(id='sigma_12',
                                        name=['sigma_12'],
                                        type='numeric',
                                        format=Format(precision=3,
                                                    scheme=Scheme.fixed)),
                                    dict(id='sigma_x',
                                        name=['sigma_x'],
                                        type='numeric',
                                        format=Format(precision=3,
                                                    scheme=Scheme.fixed)),
                                    dict(id='sigma_y',
                                        name=['sigma_y'],
                                        type='numeric',
                                        format=Format(precision=3,
                                                    scheme=Scheme.fixed)),
                                    dict(id='sigma_xy',
                                        name=['sigma_xy'],
                                        type='numeric',
                                        format=Format(precision=3,
                                                    scheme=Scheme.fixed))
```

```
                                ],
                                style_header={
                                    'textAlign': 'center'
                                },
                                style_cell={
                                    'textAlign': 'center'
                                },
                                style_data={
                                    'color': 'black',
                                    'backgroundColor': 'white'
                                },
                                style_data_conditional=[
                                    {
                                        'if': {
                                            # 'row_index': 'odd'
                                            'filter_query': '{Position} eq "Bottom"',
                                        },
                                        'backgroundColor': 'rgb(220, 220, 220)',
                                    },
                                    {
                                        "if": {"state": "selected"}, # 'active' | 'selected'
                                        "backgroundColor": "#c8f4ff",
                                        "border": "1px solid darkblue",
                                    },
                                ],
                                export_format='xlsx',
                                export_headers='ids',
                                id='tbl'
                            ),
                            title="Stress Table"
                        )
                    ],

                    start_collapsed=True,
                )
            ),

        ])


def getStressPlots(laminateObject, stressPlotData={}):
    if stressPlotData:
        stress12Values = stressPlotData["stress12Values"]
        stressXYValues = stressPlotData["stressXYValues"]
    else:
        stress12Values = getStress12Values(laminateObject)
        stressXYValues = getStressXYValues(laminateObject)
    yValues = getYValues(laminateObject)
    for index in np.arange(0, 3):
        stress12Values[index].insert(0, 0.0)
        stressXYValues[index].insert(0, 0.0)
        stress12Values[index].append(0.0)
        stressXYValues[index].append(0.0)
```

```python

    yValues.insert(0, yValues[0])
    yValues.append(yValues[-1])

    stressData = {
        "stress121": {
            "xValues": stress12Values[0],
            "yValues": yValues,
            "type": r"$\sigma_{\textit{1}}␣\text{(MPa)}$"
        },
        "stress122": {
            "xValues": stress12Values[1],
            "yValues": yValues,
            "type": r"$\sigma_{\textit{2}}␣\text{(MPa)}$"
        },
        "stress123": {
            "xValues": stress12Values[2],
            "yValues": yValues,
            "type": r"$\sigma_{\textit{12}}␣\text{(MPa)}$"
        },
        "stressXY1": {
            "xValues": stressXYValues[0],
            "yValues": yValues,
            "type": r"$\sigma_{x}␣\text{(MPa)}$"
        },
        "stressXY2": {
            "xValues": stressXYValues[1],
            "yValues": yValues,
            "type": r"$\sigma_{y}␣\text{(MPa)}$"
        },
        "stressXY3": {
            "xValues": stressXYValues[2],
            "yValues": yValues,
            "type": r"$\sigma_{xy}␣\text{(MPa)}$"
        }
    }

    stresses121 = makePlot(stressData["stress121"])
    stresses122 = makePlot(stressData["stress122"])
    stresses123 = makePlot(stressData["stress123"])
    stressesXY1 = makePlot(stressData["stressXY1"])
    stressesXY2 = makePlot(stressData["stressXY2"])
    stressesXY3 = makePlot(stressData["stressXY3"])

    return {"stresses12": {"stress1": stresses121, "stress2":
        ↪ stresses122, "stress12": stresses123},
            "stressesXY": {"stressX": stressesXY1, "stressY":
                ↪ stressesXY2, "stressXY": stressesXY3}}


def getStressXYValues(laminateObject):
    stressXYValues = [[], [], []]
    for laminaeNumber in np.arange(0, laminateObject.
        ↪ getNumberOfLaminateLayers()):
        currentLaminae = laminateObject.getLaminateLayer(laminaeNumber)
        positionArray = getPositionArray(currentLaminae)
```

```
        for yValue in positionArray:
            stressAtY = laminateObject.getStressesInXY(yValue).tolist()
            for index in np.arange(0, 3):
                stressXYValues[index].append(stressAtY[index][0])
    return stressXYValues


def getTableDataFrame(columnsValues, columnLabels):
    dataFrame = pd.DataFrame()
    for column in columnsValues:
        dataFrame = pd.concat([dataFrame, column], axis=1)
    dataFrame.columns = columnLabels
    return dataFrame.to_dict('records')


def getUnidirectionalLaminateContainer(laminateObject):
    firstLaminae = copy.deepcopy(laminateObject.getLaminateLayer(0))
    E1 = firstLaminae.getProperties()["E1"]
    G12 = firstLaminae.getProperties()["G12"]
    alphaValues = np.arange(0, 360, 1).tolist()

    eXe1PlotValues = []
    eYe1PlotValues = []
    gXYg12PlotValues = []
    v12PlotValues = []
    minusNXPlotValues = []
    minusNYPlotValues = []

    for alpha in alphaValues:
        firstLaminae.setProperties("alpha", alpha)
        complianceMatrix = firstLaminae.getTransformedComplianceMatrix()
        eXe1PlotValues.append(1 / (E1 * complianceMatrix[0][0]))
        eYe1PlotValues.append(1 / (E1 * complianceMatrix[1][1]))
        gXYg12PlotValues.append(1 / (G12 * complianceMatrix[2][2]))
        v12PlotValues.append((-1 / complianceMatrix[0][0]) *
            ↪ complianceMatrix[0][1])
        sin = np.sin(alpha * np.pi / 180)
        cos = np.cos(alpha * np.pi / 180)
        temp = (1229 * cos * sin ** 3 + 4671 * cos ** 3 * sin) / (
                3225 * sin ** 4 + 5221 * cos ** 2 * sin ** 2 + 275 * cos
                    ↪ ** 4)
        minusNXPlotValues.append(abs(temp))
        minusNYPlotValues.append(abs((-1 / complianceMatrix[1][1]) *
            ↪ complianceMatrix[1][2]))

    eXe1PlotValues.append(eXe1PlotValues[0])
    eYe1PlotValues.append(eXe1PlotValues[0])
    gXYg12PlotValues.append(gXYg12PlotValues[0])
    v12PlotValues.append(v12PlotValues[0])
    minusNXPlotValues.append(minusNXPlotValues[0])
    minusNYPlotValues.append(minusNYPlotValues[0])
    alphaValues.append(alphaValues[0])

    containerRows = []
    plotData = {
        "eXe1": {
```

```
            "title": r"$E_{x}/E_{1}$",
            "rValues": eXe1PlotValues,
            "thetaValues": alphaValues,
            "lineColor": "black"
        },
        "eYe1": {
            "title": r"$E_{y}/E_{1}$",
            "rValues": eYe1PlotValues,
            "thetaValues": alphaValues,
            "lineColor": "darkred"
        },
        "gXYg12": {
            "title": r"$G_{xy}/G_{12}$",
            "rValues": gXYg12PlotValues,
            "thetaValues": alphaValues,
            "lineColor": "black"
        },
        "v12": {
            "title": r"$\nu_{xy}$",
            "rValues": v12PlotValues,
            "thetaValues": alphaValues,
            "lineColor": "black"
        },
        "minusNX": {
            "title": r"$\text{abs}(\eta_x)$",
            "rValues": minusNXPlotValues,
            "thetaValues": alphaValues,
            "lineColor": "black"
        },
        "minusNY": {
            "title": r"$\text{abs}(\eta_y)$",
            "rValues": minusNYPlotValues,
            "thetaValues": alphaValues,
            "lineColor": "darkred"
        }
    }

    exe1Plot = makePolarDiagram([plotData["eXe1"], plotData["eYe1"]])
    gXYg12Plot = makePolarDiagram([plotData["gXYg12"]])
    v12Plot = makePolarDiagram([plotData["v12"]])
    nPlots = makePolarDiagram([plotData["minusNX"], plotData["minusNY"
        ↪ ]])

    containerRows.append(
        dcc.Markdown('''##### Pole diagram of the elastic properties
            ↪ $$E_x / E_1$$ and $$E_y / E_1$$''',
                     mathjax=True)
    )

    containerRows.append(getCustomizedGraph(exe1Plot, {"fileName": "
        ↪ eXeYe1Plot"}))
    tableData = {
        "Angle": {
            "title": "Angle",
            "precision": 1
        },
```

```
            "eXe1": {
                "title": "E_x/E_1",
                "precision": 3
            },
            "eYe1": {
                "title": "E_y/E_1",
                "precision": 3
            }
        }
        poleDiagramValues = getTableDataFrame([pd.DataFrame(alphaValues),
                                               pd.DataFrame(eXe1PlotValues),
                                               pd.DataFrame(eYe1PlotValues)],
                                              ['Angle', 'E_x/E_1', 'E_y/E_1'])
        containerRows.append(
            getAccordionDataTable(tableData, poleDiagramValues)
        )
        containerRows.append(html.Hr())

        containerRows.append(
            dcc.Markdown('''##### Pole diagram of the elastic property $$G_{xy} / G_{12}$$''',
                         mathjax=True)
        )
        containerRows.append(getCustomizedGraph(gXYg12Plot, {"fileName": "gXYg12Plot"}))
        tableData = {
            "Angle": {
                "title": "Angle",
                "precision": 1
            },
            "eXe1": {
                "title": "G_xy/G_12",
                "precision": 3
            }
        }
        poleDiagramValues = getTableDataFrame([pd.DataFrame(alphaValues),
                                               pd.DataFrame(gXYg12PlotValues)],
                                              ['Angle', 'G_xy/G_12'])
        containerRows.append(
            getAccordionDataTable(tableData, poleDiagramValues)
        )
        containerRows.append(html.Hr())

        containerRows.append(
            dcc.Markdown('''##### Pole diagram of the elastic property $$v_{xy}$$''',
                         mathjax=True)
        )
        containerRows.append(getCustomizedGraph(v12Plot, {"fileName": "v12Plot"}))
        tableData = {
            "Angle": {
                "title": "Angle",
                "precision": 1
            },
            "eXe1": {
```

```
            "title": "ðžœ^_xy",
            "precision": 3
        }
    }
    poleDiagramValues = getTableDataFrame([pd.DataFrame(alphaValues),
                                          pd.DataFrame(v12PlotValues)],
                                         ['Angle', 'v_xy'])
    containerRows.append(
        getAccordionDataTable(tableData, poleDiagramValues)
    )
    containerRows.append(html.Hr())
    containerRows.append(
        dcc.Markdown('''##### Pole diagram of the elastic properties abs($$\eta_{x}$$) and abs($$\eta_{y}$$)''',
                     mathjax=True)
    )
    containerRows.append(getCustomizedGraph(nPlots, {"fileName": "etaPlots"}))
    tableData = {
        "Angle": {
            "title": "Angle",
            "precision": 1
        },
        "eXe1": {
            "title": "abs(eta_x)",
            "precision": 3
        },
        "eYe1": {
            "title": "abs(eta_y)",
            "precision": 3
        }
    }
    poleDiagramValues = getTableDataFrame([pd.DataFrame(alphaValues),
                                          pd.DataFrame(minusNXPlotValues),
                                          pd.DataFrame(minusNYPlotValues)],
                                         ['Angle', 'abs(eta_x)', 'abs(eta_y)'])
    containerRows.append(
        getAccordionDataTable(tableData, poleDiagramValues)
    )
    containerRows.append(html.Hr())

    containerRows.append(
        dcc.Markdown('''##### Elastic constants as a function of the off-axis angle''',
                     mathjax=True)
    )

    diagramElasticProperties = {
        "eXe1": {
            "xValues": alphaValues[0:91],
            "yValues": eXe1PlotValues[0:91],
            "title": r"$E_{x}/E_{1}$",
            "lineColor": "black",
            "lineWidth": 1.5,
```

```
            "mode": "lines"
        },
        "eYe1": {
            "xValues": alphaValues[0:91],
            "yValues": eYe1PlotValues[0:91],
            "title": r"$E_{y}/E_{1}$",
            "lineColor": "darkgreen",
            "lineWidth": 1.5,
            "mode": "lines"
        },
        "gXYg12": {
            "xValues": alphaValues[0:91],
            "yValues": gXYg12PlotValues[0:91],
            "title": r"$G_{xy}/G_{12}$",
            "lineColor": "darkred",
            "lineWidth": 1.5,
            "secondaryY": True,
            "yMin": 0,
            "yMax": max(gXYg12PlotValues[0:91]) * 1.1,
            "mode": "lines"
        },
        "plotXMax": 90,
        "plotYMax": max(max(eXe1PlotValues[0:91]), max(eYe1PlotValues
            ↪ [0:91])) * 1.1,
        "xType": r"$\text{Off-axis␣angle}␣\␣\alpha$",
        "yType": r"$\text{Elastic␣constants}␣\␣E_{x}/E_{1},␣\␣
            ↪ E_{y}/E_{1}␣␣\text{␣in␣-}$",
        "secondaryYType": r"$\text{Elastic␣constant}␣\␣↪ G_{xy}/G_{12}␣
            \text{␣in␣-}$",
    }

    containerRows.append(
        dmc.Center(
            getCustomizedGraph(makeLinePlot(diagramElasticProperties), {
                ↪ "fileName": "diagramElasticPropertiesPlot1"}))
    )

    tableData = {
        "Angle": {
            "title": "Angle",
            "precision": 1
        },
        "ExE1": {
            "title": "E_x/E_1",
            "precision": 3
        },
        "EyE1": {
            "title": "E_y/E_1",
            "precision": 3
        },
        "GxyG12": {
            "title": "G_xy/G_12",
            "precision": 3
        }
    }

```

```
poleDiagramValues = getTableDataFrame([pd.DataFrame(alphaValues[0:91]),
                                        pd.DataFrame(eXe1PlotValues[0:91]),
                                        pd.DataFrame(eYe1PlotValues[0:91]),
                                        pd.DataFrame(gXYg12PlotValues[0:91])],
                                       ['Angle', 'E_x/E_1', 'E_y/E_1', 'G_xy/G_12'])
containerRows.append(
    getAccordionDataTable(tableData, poleDiagramValues)
)
containerRows.append(html.Hr())

containerRows.append(
    dcc.Markdown('''##### Elastic constants as a function of the off-axis angle''',
                 mathjax=True)
)

diagramElasticProperties = {
    "eXe1": {
        "xValues": alphaValues[0:91],
        "yValues": v12PlotValues[0:91],
        "title": r"$\nu_{xy}$",
        "lineColor": "black",
        "lineWidth": 1.5,
        "mode": "lines"
    },
    "eYe1": {
        "xValues": alphaValues[0:91],
        "yValues": minusNXPlotValues[0:91],
        "title": r"$\text{abs}(\eta_x)$",
        "lineColor": "darkgreen",
        "lineWidth": 1.5,
        "secondaryY": True,
        "yMin": 0,
        "yMax": max(max(minusNXPlotValues[0:91]), max(minusNYPlotValues[0:91])) * 1.1,
        "mode": "lines"
    },
    "gXYg12": {
        "xValues": alphaValues[0:91],
        "yValues": minusNYPlotValues[0:91],
        "title": r"$\text{abs}(\eta_y)$",
        "lineColor": "darkred",
        "lineWidth": 1.5,
        "secondaryY": True,
        "yMin": 0,
        "yMax": max(max(minusNXPlotValues[0:91]), max(minusNYPlotValues[0:91])) * 1.1,
        "mode": "lines"
    },
    "plotXMax": 90,
    "plotYMax": max(v12PlotValues[0:91]) * 1.1,
```

```
        "xType": r"$\text{Off-axis angle} \ \alpha$",
        "yType": r"$\text{Elastic constant} \ \nu_{xy}  \text{ in -}$",
        "secondaryYType": r"$\text{Elastic constants} \ \text{abs}(\eta_x), \text{abs}(\eta_y) \text{ in -}$",
    }

    containerRows.append(
        dmc.Center(
            getCustomizedGraph(makeLinePlot(diagramElasticProperties), {"fileName": "diagramElasticPropertiesPlot2"}))
    )

    tableData = {
        "Angle": {
            "title": "Angle",
            "precision": 1
        },
        "v12": {
            "title": "v_xy",
            "precision": 3
        },
        "abs(nx)": {
            "title": "abs(eta_x)",
            "precision": 3
        },
        "abs(ny)": {
            "title": "abs(eta_y)",
            "precision": 3
        },
    }

    poleDiagramValues = getTableDataFrame([pd.DataFrame(alphaValues[0:91]),
                                           pd.DataFrame(v12PlotValues[0:91]),
                                           pd.DataFrame(minusNXPlotValues[0:91]),
                                           pd.DataFrame(minusNYPlotValues[0:91])],
                                          ['Angle', 'v_xy', 'abs(eta_x)', 'abs(eta_y)'])
    containerRows.append(
        getAccordionDataTable(tableData, poleDiagramValues)
    )
    containerRows.append(html.Hr())

    return dbc.Container(containerRows)


def getUploadButtons():
    return dcc.Upload(
        id='datatable-upload',
        children=html.Div([
            'Drag and Drop or ',
            html.A('select an Excel file'),
```

```python
            ' to load the model data'
        ]),
        style={
            'width': '800px', 'height': '60px', 'lineHeight': '60px',
            'borderWidth': '1px', 'borderStyle': 'dashed',
            'borderRadius': '5px', 'textAlign': 'center', 'margin': '10px'
        },
        accept=".xlsx"
    )


def getYValues(laminateObject):
    yValues = []
    for laminaeNumber in np.arange(0, laminateObject.getNumberOfLaminateLayers()):
        currentLaminae = laminateObject.getLaminateLayer(laminaeNumber)
        positionArray = getPositionArray(currentLaminae)
        for yValue in positionArray:
            yValues.append(yValue)
    return yValues


def makeLinePlot(plotData):
    plot = make_subplots(specs=[[{"secondary_y": True}]])

    plotXMin = plotData["plotXMin"] if "plotXMin" in plotData else 0
    plotXMax = plotData["plotXMax"] if "plotXMax" in plotData else 5000.0
    plotYMin = plotData["plotYMin"] if "plotYMin" in plotData else 0
    plotYMax = plotData["plotYMax"] if "plotYMax" in plotData else 5000.0

    for subPlot in plotData.keys():
        if isinstance(plotData[subPlot], dict):
            if "secondaryY" in plotData[subPlot]:
                plot.add_scatter(
                    x=plotData[subPlot]["xValues"],
                    y=plotData[subPlot]["yValues"],
                    name=plotData[subPlot]["title"],
                    mode=plotData[subPlot]["mode"] if "mode" in plotData[subPlot] else "lines+markers",
                    marker=dict(size=plotData[subPlot]["markerSize"] if "markerSize" in plotData[subPlot] else 12),
                    line=dict(color=plotData[subPlot]["lineColor"] if "lineColor" in plotData[subPlot] else 'darkgreen',
                              width=plotData[subPlot]["lineWidth"] if "lineWidth" in plotData[subPlot] else 3),
                    secondary_y=True,
                    showlegend=True,
                )
                plot.update_yaxes(
                    range=[plotData[subPlot]["yMin"],
                           plotData[subPlot]["yMax"]],
```

```
                    gridcolor='#c0c0c0',
                    title_text=plotData["secondaryYType"],
                    secondary_y=True)
            else:
                plot.add_scatter(
                    x=plotData[subPlot]["xValues"],
                    y=plotData[subPlot]["yValues"],
                    name=plotData[subPlot]["title"],
                    mode=plotData[subPlot]["mode"] if "mode" in plotData[
                        ↪ subPlot] else "lines+markers",
                    marker=dict(size=plotData[subPlot]["markerSize"] if "
                        ↪ markerSize" in plotData[subPlot] else 12),
                    line=dict(color=plotData[subPlot]["lineColor"] if "
                        ↪ lineColor" in plotData[subPlot] else '
                        ↪ darkgreen',
                              width=plotData[subPlot]["lineWidth"] if "
                                  ↪ lineWidth" in plotData[subPlot] else
                                  ↪ 3),
                    secondary_y=False,
                    showlegend=True,
                )

    # add axes with arrow heads
    plot.add_annotation(
        x=plotXMax, # arrows' head
        y=0, # arrows' head
        ax=plotXMin, # arrows' tail
        ay=0, # arrows' tail
        xref='x',
        yref='y',
        axref='x',
        ayref='y',
        text='', # if you want only the arrow
        showarrow=True,
        arrowhead=3,
        arrowsize=1,
        arrowwidth=1,
        arrowcolor='black'
    )
    plot.add_annotation(
        x=0, # arrows' head
        y=plotYMax, # arrows' head
        ax=0, # arrows' tail
        ay=plotYMin, # arrows' tail
        xref='x',
        yref='y',
        axref='x',
        ayref='y',
        text='', # if you want only the arrow
        showarrow=True,
        arrowhead=3,
        arrowsize=1,
        arrowwidth=1,
        arrowcolor='black'
    )
    # plot.update_layout(
```

```
    # template=None,
    # )
    plot.update_layout(
        {
            # 'title': plotData["title"],
            'template': None,
            # 'plot_bgcolor': '#f3f6f4',
            'yaxis': {
                'title': plotData["yType"] if "yType" in plotData else " - "
            },
            'xaxis': {
                'title': plotData["xType"] if "xType" in plotData else " - "
            },
            'height': plotData["height"] if "height" in plotData else 550,
            'width': plotData["width"] if "width" in plotData else 850
        }
    )
    plot.update_xaxes(range=[plotXMin, plotXMax],
                      gridcolor='#c0c0c0')
    plot.update_yaxes(range=[plotYMin, plotYMax], gridcolor='#c0c0c0', secondary_y=False)

    if "logY" in plotData:
        if plotData["logY"]:
            plot.update_yaxes(type="log")

    if "logX" in plotData:
        if plotData["logX"]:
            plot.update_xaxes(type="log")

    # plot.update_traces(marker=dict(size=12))

    return plot


def makePlot(plotData):
    axisTolerance = 1.0e-12
    plot = go.Figure(
        go.Scatter(
            x=plotData["xValues"],
            y=plotData["yValues"],
            mode="lines",
            line=dict(color='black', width=2.5),
            showlegend=False,
        )
    )
    plotMargin = 0.1 # 10%

    plotXMin = -max([abs(x) for x in plotData["xValues"]]) * (1.0 + plotMargin)
    if -axisTolerance < plotXMin < axisTolerance: plotXMin = -1.0
    plotXMax = max([abs(x) for x in plotData["xValues"]]) * (1.0 + plotMargin)
    if -axisTolerance < plotXMax < axisTolerance: plotXMax = 1.0
```

```
    plotYMin = min(plotData["yValues"]) * (1.0 + plotMargin)
    plotYMax = max(plotData["yValues"]) * (1.0 + plotMargin)
    # add axes with arrow heads
    plot.add_annotation(
        x=plotXMax, # arrows' head
        y=0, # arrows' head
        ax=plotXMin, # arrows' tail
        ay=0, # arrows' tail
        xref='x',
        yref='y',
        axref='x',
        ayref='y',
        text='', # if you want only the arrow
        showarrow=True,
        arrowhead=3,
        arrowsize=1,
        arrowwidth=1,
        arrowcolor='black'
    )
    plot.add_annotation(
        x=0, # arrows' head
        y=plotYMax, # arrows' head
        ax=0, # arrows' tail
        ay=plotYMin, # arrows' tail
        xref='x',
        yref='y',
        axref='x',
        ayref='y',
        text='', # if you want only the arrow
        showarrow=True,
        arrowhead=3,
        arrowsize=1,
        arrowwidth=1,
        arrowcolor='black'
    )
    plot.update_layout(
        {
            # 'title': plotData["title"],
            # 'plot_bgcolor': '#f3f6f4',
            'template': None,
            'yaxis': {
                'title': r'$z␣\text{(mm)}$'
            },
            'xaxis': {
                'title': plotData["type"]
            },
            'height': 500
        }
    )
    plot.update_xaxes(range=[plotXMin, plotXMax], gridcolor='#D3D3D3')
    plot.update_yaxes(range=[plotYMin, plotYMax], gridcolor='#D3D3D3')

    return plot


def makePolarDiagram(plotData):
```

```
    plot = go.Figure()

    for subPlot in plotData:
        plot.add_scatterpolar(
            r=subPlot["rValues"],
            theta=subPlot["thetaValues"],
            mode="lines",
            name=subPlot["title"],
            # line=dict(color='darkgreen', width=2),
            showlegend=True,
            line=dict(color=subPlot["lineColor"] if "lineColor" in subPlot else 'darkgreen',
                      width=subPlot["lineWidth"] if "lineWidth" in subPlot else 1.5),
        )

    plot.update_traces(
        hovertemplate=None
    )
    plot.update_layout(
        template=None,
        height=550
    )
    plot.update_layout(
        template=None,
        polar=dict(radialaxis=dict(gridwidth=0.5,
                                   # range=[0, 3],
                                   range=[0, max(plotData[0]["rValues"]) * 1.5],
                                   showticklabels=True, ticks='', gridcolor="lightgrey"),
                   angularaxis=dict(showticklabels=True, ticks='',
                                    rotation=0,
                                    direction="counterclockwise",
                                    gridcolor="grey")
                   )
    )
    return plot


def makeStressStrainPlot(plotData):
    plot = go.Figure(
        go.Scatter(
            x=plotData["xValues"],
            y=plotData["yValues"],
            mode="lines+markers",
            line=dict(color='darkgreen', width=3),
            showlegend=False,
        )
    )
    plotMargin = 0.1 # 10%
    plotXMin = 0
    plotXMax = max(plotData["xValues"]) * (1.0 + plotMargin)
    plotYMin = 0
    plotYMax = max(plotData["yValues"]) * (1.0 + plotMargin)
    # add axes with arrow heads
```

```
    plot.add_annotation(
        x=plotXMax, # arrows' head
        y=0, # arrows' head
        ax=plotXMin, # arrows' tail
        ay=0, # arrows' tail
        xref='x',
        yref='y',
        axref='x',
        ayref='y',
        text='', # if you want only the arrow
        showarrow=True,
        arrowhead=3,
        arrowsize=1,
        arrowwidth=1,
        arrowcolor='black'
    )
    plot.add_annotation(
        x=0, # arrows' head
        y=plotYMax, # arrows' head
        ax=0, # arrows' tail
        ay=plotYMin, # arrows' tail
        xref='x',
        yref='y',
        axref='x',
        ayref='y',
        text='', # if you want only the arrow
        showarrow=True,
        arrowhead=3,
        arrowsize=1,
        arrowwidth=1,
        arrowcolor='black'
    )
    plot.update_layout(
        {
            # 'title': plotData["title"],
            'plot_bgcolor': '#f3f6f4',
            'yaxis': {
                'title': plotData["yType"]
            },
            'xaxis': {
                'title': plotData["xType"]
            },
            'height': 500
        }
    )
    plot.update_xaxes(range=[plotXMin, plotXMax], gridcolor='#c0c0c0')
    plot.update_yaxes(range=[plotYMin, plotYMax], gridcolor='#c0c0c0')
    plot.update_traces(marker=dict(size=12,
                                   line=dict(width=2,
                                             color='DarkSlateGrey')))

    return plot


def parse_contents(contents, filename):
    content_type, content_string = contents.split(',')
```

```
    decoded = base64.b64decode(content_string)
    if 'xls' in filename:
        # Assume that the user uploaded an Excel file
        return [pd.read_excel(io.BytesIO(decoded), sheet_name="
            ↪ ModelProperties"),
                pd.read_excel(io.BytesIO(decoded), sheet_name="LoadMatrix
                    ↪ "),
                pd.read_excel(io.BytesIO(decoded), sheet_name="
                    ↪ thicknessMatrix")]


def updateLayerColumn(rows):
    counter = 0
    for i in np.arange(0, len(rows)):
        counter += 1
        rows[i]["Layer"] = counter
    return rows
```

5.5 Requirements.txt

requirements.txt (requirements)

```
dash>=2.6.0
dash_bootstrap_components>=1.0.3
dash_mantine_components>=0.11.1
numpy>=1.22.2
pandas>=1.4.1
plotly>=5.6.0
sympy>=1.10.1
XlsxWriter>=3.0.3
openpyxl>=3.0.10
```

Index

A. Öchsner and R. Makvandi, *A Numerical Approach to the Classical Laminate Theory of Composite Materials*, Advanced Structured Materials 189,
https://doi.org/10.1007/978-3-031-32975-3

Lecture Notes in Mechanical Engineering

ABCM Series on Mechanical Sciences and Engineering

This series publishes selected papers as well as full proceedings of events organized and/or promoted by the Brazilian Society for Mechanical Sciences and Engineering (ABCM) on an international level. These include the International Congress of Mechanical Engineering (COBEM) and the International Symposium on Dynamic Problems of Mechanics (DINAME), among others.